9504412
LBGR

BIRMINGHAM CITY
UNIVERSITY
DISCARDED

Global Construction and the Environment

GLOBAL CONSTRUCTION AND THE ENVIRONMENT

STRATEGIES AND OPPORTUNITIES

FRED MOAVENZADEH

A WILEY-INTERSCIENCE PUBLICATION
JOHN WILEY & SONS, INC.
NEW YORK / CHICHESTER / TORONTO / BRISBANE / SINGAPORE

This text is printed on acid-free paper.

Copyright © 1994 by John Wiley & Sons, Inc.

All rights reserved. Published simultaneously in Canada.

Reproduction or translation of any part of this work beyond
that permitted by Section 107 or 108 of the 1976 United
States Copyright Act without the permission of the copyright
owner is unlawful. Requests for permission or further
information should be addressed to the Permissions Department,
John Wiley & sons, Inc. 605 Third Avenue, New York, NY
10158-0012.

This publication is designed to provide accurate and
authoritative information in regard to the subject
matter covered. It is sold with the understanding that
the publisher is not engaged in rendering legal, accounting,
or other professional services. If legal advice or other
expert assistance is required, the services of a competent
professional person should be sought.

Library of Congress Cataloging in Publication Data:
Moavenzadeh, Fred, 1935–
    Global construction and the environment : strategies and
 opportunities / by Fred Moavenzadeh.
      p.   cm.
    A Wiley-Interscience Publication
    Includes bibliographical references (p.   ) and index.
    ISBN 0-471-01289-0 (alk. paper)
    1. Civil engineering—Environmental aspects. 2. Construction
 industry—Environmental aspects. 3. Industrial management—
 Environmental aspects. I. Title.
TD195.C54M63   1994
363.73'1—dc20   93-6346

Printed in the United States of America

10 9 8 7 6 5 4 3 2 1

CONTENTS

PREFACE   xiii

INTRODUCTION   xvii

CHAPTER 1
ENVIRONMENT AND THE CONSTRUCTION INDUSTRY   1

Construction Firms Need to Change Their Structure	4
Tighter Regulations Shift the Onus to Construction Contractors	7
The Push Is On to Eliminate Waste at the Source	9
Teamwork Approach	11
Why Alliances Make Sense in the Environmental Marketplace	*13*
Numerous Barriers Stifle Innovation	16
What's Needed to Spur Innovation?	*17*
Better Communications on New Technologies Needed	21

CHAPTER 2
INDUSTRIAL PERSPECTIVES AND INITIATIVES   25

From Neglect to Protection	27
Environmental Pressures Build	29
Industry versus Environmentalism	30
Rising Concern about Ecology and Industry's Role	32
The Traditional Economic Model	33
Failures Frustrate Policy Makers	37
U.S. Corporations Forced to Review Environmental Policies	40
The 3M Pollution Prevention Program	42
AT&T's Process Improvements	45

Pollution Reduction in Energy Production 48
Emergence of the Era of Sustainable Development 50
The New View: Asset Management 51
Experience in Europe 52

CHAPTER 3
ENVIRONMENTAL POLICY AND REGULATIONS 57
Policy Perspectives and the Environment 61
Environmental Policy: How and Why 63
Development of Environmental Regulation in the United States 64
Command-and-Control Regulation 65
The Case for Market-Based Incentives 68
    Typical Market-Based Programs: Advantages and Disadvantages 70
Difficulties of Implementing Market Mechanisms 75
Citizen Empowerment and the Right to Know 77
Diversity in Perspective on Environmental Protection 78
Major Areas of U.S. Policy and Regulation 79
    Airborne Waste 80
    Waterborne Waste 81
    Solid and Hazardous Waste 82
Learning from Foreign Experiences 85
International Accords 89

CHAPTER 4
TECHNOLOGICAL CHALLENGES AND OPPORTUNITIES 95
Regulation Drives the Environmental Market 97
How Traditional Types of Regulation Impact Technological Development 99
Market-Incentive Regulations 101
Environmental Progress: Salience of Trade-Offs 103

Technologies Follow Directions Set by Policies   103
Hazardous Waste   105
  In-Situ Systems   *106*
  Prepared-Bed Systems   *106*
  In-Tank Systems   *106*
Solid Waste   113
  Sanitary Landfill   *114*
  Incineration   *115*
  Recycling and Composting   *118*
  Waste Minimization   *119*
Technology Tackles Energy Problems   120
Energy Alternatives to Fossil Fuels   124
Battling the Barriers to Innovation   125

CHAPTER 5
ENVIRONMENTAL MARKETS   **133**
How Market Demand Is Created   134
Dynamics of the Environmental Marketplace   136
Structure of the Environmental Marketplace   137
Major Growth in U.S. Environmental Markets   140
Opportunities in Environmental Market Sectors   142
  Environmental Assessment   *142*
  Hazardous Waste   *143*
  Solid Waste   *147*
  Waste Minimization   *153*
  Sewer/Wastewater   *154*
  Water Supply   *155*
  Air Pollution Abatement   *156*
Regulatory Impacts on Energy and Petrochemical Industries   157
  Energy   *158*
  Petrochemicals   *163*

Worldwide Environmental Market Trends 165
   European Markets 165
   Pacific Rim Markets 169

CHAPTER 6
SHIFTING ATTITUDES CREATE OPPORTUNITIES FOR CONSTRUCTION — **173**

Survey Questionnaire Design 176
   Section I: General Information about Your Firm 176
   Section II: The Environmental Market 176
   Section III: Response to the Environmental Market 176

Characteristics of Survey Respondents 177
The Construction Industry: *The Supply Side* 178
Owners of Constructed Facilities: *The Demand Side* 182
Perspectives on the Environmental Markets 182
   The Construction Industry: The Supply Side 183
   Owners of Constructed Facilities: The Demand Side 192
Business Strategies 197
   The Construction Industry: The Supply Side 197
   Owners of Constructed Facilities: The Demand Side 199
Environmental Awareness and Responsibility 201
The Construction Industry: *The Supply Side* 202
Owners of Constructed Facilities: *The Demand Side* 205

CHAPTER 7
CORPORATE STRATEGIES AND FUTURE TRENDS — **209**

Dealing with Risk, Liabilities 212
Strategic Vision Needed 216
Preserving and Expanding Wealth 217
Placing Values on the Environment 218

From Regulations to Incentives	220
The Opportunities	223
APPENDIX A	**225**
MIT Symposium Program	225
MIT Symposium Summaries	000
APPENDIX B	**241**
Questionnaires for Industry Surveys	000
Survey Questionnaire	242
Survey Questionnaire (Owner)	000
REFERENCES	**281**

PREFACE

To explore the potential for positive contributions by the construction industry to concerns over the global environment, and to help identify market opportunities for construction firms in this rapidly expanding field, a Symposium on the "Global Environment and the Construction Industry" was held October 21–22, 1991, at the Massachusetts Institute of Technology in Cambridge, Massachusetts. The Symposium was sponsored jointly by the Hazama Corporation, one of Japan's leading construction companies, and the Center for Construction Research and Education (CCRE) at MIT.

This Symposium brought together representatives from industry, government agencies including the EPA, the military services, construction firms, and academic institutions. The meeting provided a forum for discussion and the exchange of ideas on environmental concerns, policies and regulations, and future directions in technology and the marketplace. One goal of the Symposium was to provide a forum for balanced discourse; to help illuminate some of the many gray areas surrounding the clash between ecology and economic development; and to identify the role that construction firms can play in the rapidly growing environmental marketplace.

Using presentations from this conference as a base, I will explore several major aspects of the environmental movement and its interaction with the construction industry in the ensuing pages. I have consulted numerous additional sources to add meaningful input to topics touched on in meeting discussions. Summaries of the Symposium presentations appear in Appendix A, and additional sources are identified in the Preface and References.

Robert Haavind, a business/technology writer and consultant, contributed significantly to the writing of the manuscript. Important contributions came from MIT graduate students and were based on their theses which have been published by the CCRE, Department of Civil and Environmental Engineering at MIT as part of a series of reports on Global Environment and the Construction Industry. These include:

- The Hazardous Waste Remediation Market: Innovative Technological Development and the Growing Involvement of the Construction Industry, by Andrew J. Hoffman

- Solid Waste Management: Decision and Market Dilemmas, by Henry F. Taylor III
- Future Construction Demands in Environmental Markets, by Edmund S. Pendleton
- Construction Opportunities in Environmental Markets, by Edmund S. Pendleton
- Valuable feedback and suggestions on the text were also provided by Hoffman and Pendleton.

The CCRE was formed in the Department of Civil and Environmental Engineering in 1982. The mission of the Center is threefold: (1) education, (2) research, and (3) industrial involvement. Its objectives are to:

- provide a research environment conducive to development and application of innovative construction technologies and management principles
- offer graduate programs in construction engineering and management
- act as a facilitator and catalyst for improving the productivity and competitiveness of the engineering and construction industry and to enhance its contribution to the economy and society in general

The Center's academic mission is implemented through a graduate-degree program in Construction Engineering and Management. This program has nearly 200 alumni worldwide. The CCRE has about 50 graduate students—one-third at the Ph.D. level and two-thirds at the Masters level. The majority of its students take at least 25 to 75 percent of their subjects in the area of management.

The objectives of the research program at CCRE are to discover, synthesize, and apply knowledge needed to promote fundamental, long-term improvements in U.S. construction productivity and capability. The Center's efforts have been focused primarily in the following areas:

- Research on *advanced materials* concentrates on potential applications of new engineered materials such as ceramics, composites, cellular materials, and foam-like materials.
- In the area of *information technology,* the Center's focus is on both hardware (automation, robotics, remote sensing, smart tools) and on Software (computer-aided design and construction, and computer-integrated construction).
- Efforts in *technology evaluation* focus on the nature of knowledge transfer in the construction industry. The general thrust is to determine empirically the actual mechanisms of technological development and acquisition, and

thereby build a better basis for a strategy of innovation and transfer of technology.
- Investigations of the *infrastructure technology* seek to understand the manmade, natural, and social environments that combine to shape our national infrastructure. By increasing our knowledge of their performance, the Center aims to improve the efficiency of managing and maintaining these complex, large-scale systems.
- The Center's efforts in the *management of technology* focus on construction management, project management, firm operations, construction finance, strategic management, and policy issues for the construction industry at the regional, national, and global level.

The Center has recently formed a Consortium on the Construction Industry and Global Environment, which provides a research atmosphere for defining and fostering a comprehensive understanding of critical issues, identifying opportunities for new markets, appraising emerging technologies, and assisting member firms in developing appropirate strategies for participating in new ways to mitigate or avoid ecological damage to the environment.

A vital component of CCRE's overall mission is to transfer the knowledge, experience, and research results that it has achieved to industry. To do so, it has developed a partnership among government, academia, and industry. The Center encourages exchange of personnel to interact and work with their respective counterparts on research projects. The primary mechanism for bringing industry knowledge to its program is to invite industry members to join the Center for various periods of time as senior lecturers or visiting engineers.

The Center disseminates its findings through research reports and articles, papers, and student theses. In addition, it publishes a newsletter three times a year that is distributed free of charge to members of industry.

Fred Moavenzadeh
Director
Center for Construction
Research and Education

INTRODUCTION

INTRODUCTION

The construction industry has a special place in the development of human history and has played an important part in shaping society's physical environment. Construction firms are responsible for planning, designing, and constructing the buildings used for commerce and shelter, as well as for all the infrastructure and facilities needed to provide basic services, such as energy and water, along with the transportation and communications systems that allow modern society to function. The industry participates in every phase of development, from investment and financing to site planning engineering, and architecture; through project execution; and even on into facilities management. The industry's activities range from massive projects, including petrochemical and industrial plants, airports, skyscrapers, highways, bridges, and dams, to smaller jobs such as office and apartment buildings or single-family homes. It is clear, then, that the construction industry plays a central role in economic development.

Construction, while carrying out its mission, has made major, long-lasting alterations to our natural environment. Construction is, by definition, the modification of nature. Constructors, in effect, modify the natural world to make it more suitable for a great variety of human activities. Still, until the past couple of decades, the environmental consequences of construction activities went largely unnoticed—even though performing their tasks put builders directly at the interface between human/social needs and the environment.

Although builders have long been accustomed to dealing with environmental issues, the intensifying debate between environmentalists and developers in recent times has regulated in ever increasing constraints on construction projects of all types. The trend will undoubtedly continue, so it is important for those in the construction industry to be aware of the roots and the dimensions of the environmental movement. It is important as well for anyone concerned about the environment to understand construction's role. This knowledge can help construction firms understand and comply with current regulatory and permit requirements; it can also help industry executives to prepare for future markets that will be created and shaped by more stringent environmental requirements. Better understanding of the construction process can, in turn, help those interested in environmental preservation to be realistic in shaping demands on this critical industry.

Even though stepped-up environmental requirements mean new constraints for the industry, they also open up new opportunities. Those construction firms knowledgeable about trends in the environmental field will be better positioned to take advantage of emerging needs in this fast-growing marketplace. A first important step for these companies is to recognize that the environmental sector differs significantly from traditional construction markets. Understanding the unusual characteristics of this fledgling industry is a prerequisite for firms expecting to capitalize on new opportunities. It is also important to recognize the wide range of market possibilities for many of the skills and capabilities already abundant in the construction business. For example, cleaning up the environmental degradation resulting from practices of the past calls for implementation of a wide range of traditional construction services, including site planning, project management, earth moving, and transportation of contaminated materials. However, it also calls for new capabilities in decontamination technologies, either off-site or on-site, as well as specialized techniques for treatment of toxic substances and prevention of groundwater contamination.

To become active participants in this unique and complex market requires knowledge not just about regulations and licensing but also about a range of other environmental topics. This additional set of insights will help the industry to better serve its existing clients and as the same time identify new business opportunities. Clients will expect their construction contractors to be familiar with emerging environmental technology so that they can help them deal with environmental regulations more efficiently. Industrial customers will increasingly be seeking assistance from the construction industry in finding ways to reduce or eliminate waste at the source. All of these rising expectations make it vital to be aware of important trends in environmental policy, regulations, technology, and markets.

This book is written to give construction industry managers and decision makers insight into the environmental movement and its impact on their businesses. It therefore discusses important environmental issues, but it also describes new technologies, analyzes market segments that offer business opportunities, and probes the shifts in attitudes and objectives that will shape future environmental policies pertinent to construction activities. The book is designed to give construction executives an overview of the impact of the growing global environmental movement, bringing them up to date on some major issues including those involving environmental regulations and policies. This understanding can provide a framework for developing plans and strategies not just for responding to environmental pressures, but also for active and profitable participation in an important and fast-growing marketplace. The book also will help others involved in global environmental change to appreciate the complex interactions between construction activities and the natural world.

ENVIRONMENTAL "CRISIS": DANGER AND OPPORTUNITY

Today, the environment has become the subject of great concern to many—for some it is indeed in crisis. For the construction industry, the nature of the "environmental crisis" is best illustrated by reflecting on how the Chinese represent "crisis" as a printed character. To the Chinese, there is no single word what means crisis. Instead, crisis is represented by combining two characters that, literally translated, mean "danger" and "opportunity." This combination of meanings—danger and opportunity—is exactly how the environmental movement is likely to affect the construction industry.

First, there is danger for the industry, in the form of fundamental constraints on construction projects. (Compliance with environmental regulations typically increases the costs of construction, not just for building but for many related activities, such as waste disposal and materials costs. A major public issue in the United States, for example, is the dramatic loss of old-growth forests in the Northwest due to the worldwide demand for lumber for use in building products. Restrictions may limit lumber supplies, causing prices to rise and forcing the use of substitutes. The production of other construction materials, such as cement, steel, and aluminum, can release toxins into the air. Earthworks may leave open areas susceptible to erosion, contributing to surface-water degradation. Discarded building materials and demolition refuse make up a significant portion of the debris that is rapidly filling up scarce landfill space. It is difficult, in fact, to find any aspect of construction that does not have some kind of impact on the environment. The erection of anything will at the very least change the scenery! Mitigating potential detrimental effects of these environmentally related concerns will surely raise costs for builders, at least in the short run. If the cost of construction rises, according to basic economics, *ceteris paribus*, demand should fall. Although this is a very bald argument, it does illustrate an element of the danger posed by the emergence of environmental awareness.

Even though dangers posed to the industry are real, the opportunities that will appear due to the new environmental consciousness may far overshadow the dangers. The markets that are developing to meet national and worldwide environmental needs are already quite large and are still growing quickly. They may offer potentially stable markets for those construction firms that can recognize the opportunity and adapt to market needs.

But even beyond actually entering the marketplace, construction firms (along with companies in virtually every industrial sector) will increasingly find that environmental considerations shape all of the work that they do. Actually, the construction industry has the potential to become a potent force for reducing or even eliminating many of the causes of environmental degradation. Accomplishing this will require construction companies to expand their horizons and work closely with clients to help them meet environmental challenges.

A wide range of opportunities exist for construction firms that want to contribute to environmental progress. Every phase of the development process is amenable to environmental improvement, from producing building materials to dealing with the wastes created during and after construction. Firms can factor environmental considerations into site preparation and the planning phases of new projects. Builders can also work with clients to improve the operating efficiency (and reduce the operating costs) of buildings and plants, providing owners with high-performance structures that maximize available material and energy efficiencies and minimize waste and pollution. There is a huge, growing demand in all areas of environmental services such as hazardous and nuclear waste remediation, solid waste management, air pollution control, wastewater treatment, water supply, and energy conservation. The policies employed to help solve specific environmental problems will also create new demands. The recent emphasis on recycling, for example, has produced a need for recovered-material reprocessing facilities, such as urban mini paper mills and paper de-inking plants.

SHIFTS IN PUBLIC ATTITUDES

Significant developments have occurred slowly and deliberately in the area of environmental policy and regulation during the past three decades. Environmentally troublesome accidents such as oil spills, seepage from toxic waste dumps, and acid rain consistently grab the headlines, but overall fundamental changes in governmental philosophy regarding the environment remain mostly obscure. There have indeed been significant developments in substance and direction for both environmental policy and regulations, occurring for the most part over the last two decades, and primarily due to public demand for better protection of the environment. These trends and developments are not likely to be reversed, so those developing strategies for the future should be aware of emerging policy trends that will increasingly impact all environmentally related activities.

Since 1970, the generally accepted view of the earth's capacity to absorb mankind's wastes has gone through a major reassessment. During the preceding two centuries, industry, and even society as a whole, tended to view the world as an infinite sink for human wastes and pollution. It was believed that, given time, natural processes in the atmosphere, soil, and rivers and oceans would eventually disperse and neutralize any by-products of human activities. Even though this mind-set was almost universal, it was not really supported by the facts. Throughout history, human societies have left a lengthy record of environmental destruction. But environmental degradation tended to be a slow process, concentrated in a few crowded areas where the technologies available failed to deal adequately with gradual accumulations of waste and sewage.

In more recent times, however, the growth and intensified concentration of populations, rapid industrialization, and advancing technology helped to create an exponentially increasing capacity for human activities to affect the natural world. This rising capacity for degradation, coupled with apathy about environmental consequences, led to serious hazards to health and quality of life. Citizens found themselves surrounded by polluted waters and air and threatened by nearby dumps leaking toxic chemicals into groundwater and soils. It became clear that nature could not possible cope with the burgeoning waste products of civilization and industrialization. Pressures built for action to stem the worst environmental abuses.

There were many such abuses in the relatively constraint-free era before the 1970s. It was common, for example, to dispose of toxic substances by pouring them into 55-gallon drums and burying them. Although there has since been much criticism of industry practices prevalent in that era, we should recognize that companies were often complying fully with whatever regulation existed at the time. As environmental consciousness began to rise, a variety of steps were taken during the 1970s and 1980s to temper the worst practices.

Now, in the 1990s, the whole era of environmental abuse is coming to a fairly abrupt end. The shifts that are rapidly taking shape go beyond merely dealing with the undesirable direct consequences of past practices that must be dealt with local and regionally. We now recognize that humans are part of what may be fragile ecosystems that must be adequately maintained to ensure not just health but even human existence. If we want to sustain the vitality of these ecosystems, we must gear human activities to preserve clean air, clean waters, forests, wetlands, and the plant and animal species that inhabit the earth. While these concepts have been applied to individual regions and countries, there is now growing pressure to ensure that advancing development does not disturb the health of the earth itself, such as by causing changes in global weather patterns or by disrupting the protective stratospheric ozone layer. Several accords concerning the global environment emerged from an unprecedented international conference, the Earth Summit, attended by many heads of state from all over the world. Some believe that these global agreements are the start of concerted international cooperation that will eventually prove as vital to world order as the United Nations Charter of 1945.

This dramatic change in public attitudes had humble beginnings. A conservation ethic began to arise at about the turn of the century in the United States, focused mainly on preserving some of the nation's most scenic areas in a pristine, natural state through the establishment of national parks. In the postwar period, few new major environmental steps were taken as the industrial economy began booming. Chemistry and nuclear science each promised to enhance life in new, unprecedented ways. By the end of the 1960s, it had become apparent that there were environmental costs attached to applying sciences such as these to the everyday world. In *Silent Spring,* Rachel Carson chronicled the effects of pesticides on bird populations—shocking the nation by revealing the hidden consequences associated with human attempts to control and manipulate the natural environment.

In the same period, the limitations of the earth's resources and the need for conservation became a public concern, due especially to the oil embargo of

1973. Many wondered when the effects from an ever-increasing human population would become a threat to sustaining life. This view of the limits to growth suggested that severe shortages of critical materials and energy supplies, coupled with fast-increasing amounts of waste and pollution, might overwhelm civilization within a century or so. Averting such a catastrophe would require drastic action across a broad front, radically changing existing patterns of development. Partly as a result of such fears, legislation was passed setting up what are commonly called "command-and-control" regulations aimed at curbing pollution and better managing waste.

Currently, worldwide development trends add urgency to such environmental concerns. The output of developing nations is expected to rise some 4–5 percent per year between 1990 and 2030, according to a report by the World Bank. This increase translates to a fivefold jump in all output, especially important since modernization and industrialization can cause significant environmental problems. Development in just the industrialized countries is expected to expand less rapidly, but it is still predicted to triple over this time span. By 2030 total global output would expand to 3.5 times today's level, with a threefold rise in the use of energy. If pollution and waste generation were to rise in step with this economic expansion, the impact on the environment and on human well-being could be devastating. The effects of air and water pollution would sicken or kill tens of millions more people than it already does today. There would be acute water shortages, and only a small fraction of the current tropical forests and other natural habitats would remain.

ENVIRONMENTAL POLICY AND LEGISLATION

This gradual acknowledgment of the effects that humans have on the environment has led to an increased appreciation for the complexity and fragility of the natural world. Over time, the new view has fundamentally altered the ways that government policy makers manage nature-human interaction. During the past two decades, many nations have grappled with methods to promote sounder environmental management. Nations have been trying to curb environmental abuses by defining "allowable" practices in the law.

The task has proven to be very difficult. It is not enough to isolate some practice, identify offending substances, set "acceptable" limits for emissions, require reports on emissions and discharges based on monitoring and measurements, and then force compliance. As discussions later in this book will demonstrate, each of the steps in a command-and-control regime involves complexities that may not be immediately apparent. Policy makers need to develop a more holistic view as they consider the benefits and drawbacks of growing sets of alternative policy options for controlling the practices of industry. Zealots intent

INTRODUCTION

on dominating policy making may be found on each side of debates over environmental issues. Factual analysis can help resolve such disputes in some policy areas. In the environmental arena, unfortunately, it is often not practical to resolve conflicts by simply referring to data determined by experimentation and measurements. A straightforward analysis of the state of knowledge presented by experts in a particular area of concern, even when facts are carefully considered to help evaluate options and to reach optimum choices, does not guarantee that a credible, good, or implementable decision will be reached.

Many environmental areas are either not yet well understood or are not even amenable to factual analysis. Choices must be made, frequently requiring difficult value judgments and often with inadequate information. This often subjective approach can allow decisions to be influenced by those with political and economic power. Even though stakeholders in government, industry, and citizens groups may reach general agreement on environmental goals, in practice, huge barriers often remain. This may prevent the changes in lifestyles, public attitudes, and industrial operations needed to attain the agreed-upon objectives.

For some environmental problems, the command-and-control paradigm was sufficient to achieve limited goals, but problems remained that were more persistent. Laws based on command-and-control principles tend to affect industries unevenly, causing great hardship for one firm while having minimal effect on another firm emitting the same type of pollution. Another flaw is that scarce resources might be devoted to cleaning up existing facilities to meet regulations rather than going toward modernized, more efficient plants that would eliminate many pollution and effluent problems in the first place, without so many costly control systems. Policy makers looked for ways to induce changes in industrial behavior that were more balanced and effective. Replacing the stick with a carrot through a system of incentives appeared to offer an attractive alternative.

Incentive-based policies have the potential to equalize compliance costs for industry while making pollution control more cost-effective. In cases where there is open access to resources and little incentive to conserve or replace them, degradation occurs even when environmental laws are enforced. Such is the case with forests, clean air and waters, and many of the world's fisheries, for example. While there appears to be little alternative to command-and-control regulation and enforcement in some environmental areas, incentives may prove much more effective where they are feasible.

Currently, policy makers are experimenting with regulatory approaches that build more incentives and penalties right into the basic market system in hopes of deterring, rather than trying to control, environmentally destructive practices. In sectors where market forces are at play, there tends to be much smoother, more efficient adjustment to changing conditions. Where environmental costs are reflected in market prices, such as in the markets for metals, minerals, and even energy, the forces of technical progress, substitution, and structural change can come into play to alleviate problems.

BEYOND WASTE MANAGEMENT

Just controlling pollution and treating waste is not enough to reach the ultimate ideal of "sustainable development," or finding ways to build the facilities necessary to meet human needs and desires without degrading the quality of life for present and future generations. Extensive research and development (R&D) is yielding new products and processes that minimize the use of energy and greatly reduce or sometimes even eliminate all harmful solid, liquid, or gaseous waste. In some cases, residual by-products from manufacturing processes are being turned into useful materials that can be put to work in other processes or incorporated into different products. Economic benefits from such resource recovery and reuse flow both to the developers of the enabling technologies and to the industries that employ them. Benefits could include lower materials and production costs through increased efficiency, lower disposal and treatment costs as waste is reduced or eliminated, and improved products capable of increasing market share. Source control methods also promise to offer real and lasting solutions to many nettlesome environmental problems. They can make an important contribution toward the achievement of sustainable development.

But substantive gains cannot be made by using the methods of the past. although stepped-up R&D can produce promising new environmental technologies, practical progress will depend on actually putting these advances to work. The construction industry can be instrumental in moving emerging technologies into the marketplace, and those companies most adept at doing so may gain a competitive advantage. A construction firm can use a few technologies, such as impervious-barrier landfill designs, for its own projects. Most opportunities, however, will involve incorporating new technologies into structures and facilities being built for clients in other industries.

NEW INITIATIVES IN INDUSTRY

While the regulatory framework is changing, some industrial firms have already taken their own initiative and reached beyond merely following environmental rules. Executives are recognizing that generating pollution or other waste only adds to the costs of manufacturing, even putting aside the steadily increasing costs of environmental management. Investments in environmental R&D and in new processes that reduce or eliminate waste have begun to produce attractive returns for some companies. This quest for improvements in environmental quality is expanding markets and creating new opportunities for industry in general, and specifically for construction firms. Curbing sources of pollution and waste, improving energy efficiency for industrial processes, and providing more efficient climate control within buildings are among the areas of growing demand.

A rapidly growing market is also directed at cleaning up existing hazardous waste sites. Beyond basic earthmoving capabilities, this work requires skills and technologies suitable for treating toxic materials present in soil and groundwater. New technologies that allow such work to proceed right on site and often in place (in situ) have become the subject of intensive research. These are much more cost-effective since they eliminate the need for moving large amounts of hazardous materials off-site for treatment. A growing number of construction firms are entering this burgeoning market, either by acquiring the necessary skills and licensing applicable technology, by making acquisitions of firms within chosen environmental sectors, or by forming alliances with environmental and other specialty firms such as those experienced in transporting hazardous waste.

A few U.S. firms are gaining positions of global leadership in these emerging markets, as well as in more traditional pollution-control and waste-management markets, through the development of improved technology and methods. This know-how has great potential not only for the U.S. marketplace, but also for helping address mounting problems in developing nations and in the splintered remains of the environmentally ravaged nations of the former Soviet bloc.

Just as the construction business itself is becoming globalized, the industry may find expanded opportunities to participate in the worldwide quest for environmental improvement. In many cases, alliances with foreign firms will be essential for tracking and gaining entry into these growing international markets. These opportunities are arising, particularly in Western Europe—because U.S. regulations have been more stringent, American firms have developed the project management and other skills needed for work in Europe. At the same time, it is important for American industry to recognize that some 70 percent of the environmental technology applied in the United States is now being provided by foreign companies, despite the fact that the techniques are often based on research originally done in this country. Japan and Germany are particularly effective in carving niches in the emerging environmental markets through the development and application of advanced technology.

THE ROLE OF TECHNOLOGY

Construction firms can also participate as partners in advanced technology projects with potential for future commercialization. Many opportunities exist in the energy sector, particularly in alternate energy sources that are being developed as substitutes for dwindling supplies of hydrocarbons. The construction industry will play an important role in building and installing these emerging energy-conserving or energy-producing technologies.

Rapid progress toward the sustainable development ideal will require much greater interaction between the research community and those in the construction or environmental firms charged with implementing real-world solutions. Better communication between those planning designing, building, or equipping new facilities and those working on new technologies is essential. Even with better communication, however, the difficulties in dealing with some environmental problems seem monumental.

Some industries are actively working on source-control methods, and, if suitable processes are not available on a turnkey basis, they are cobbling together alternatives of their own. A few companies are taking the lead in rewarding employees and managers who can help find constructive ways to improve environmental performance.

Government agencies are recognizing that they must find more flexible approaches to address the barriers faced by developers and implementors of new environmental technologies. Construction firms are beginning to forge alliances and develop strategies that can turn environmental markets into strong profit centers for them.

While this emerging industry may seem enticing, potential entrants must be aware of potential pitfalls. Decision makers in construction firms need to consider the dangers of environmental liability, recognizing that any mistakes in judgment or execution when designing or constructing environmentally related projects can prove extremely costly.

Aside from these cautions, however, there is potential for substantial growth, and many firms have enjoyed high profitability and unparalleled growth in environmentally related industries. Construction firms willing to face some risk while building a competitive position in this market place could be the next to prosper.

GLOBAL MARKET

Global concern over the potential ill effects of uncontrolled development and industrial activity has been rising as policy makers realize that the solutions to many environmental problems are linked to increased cooperation and require a global perspective. There has been an evolution in ideas about how individual nations and the world community can alleviate or prevent world-scale environmental problems. The United Nations Conference on Environment and Development in Rio de Janeiro in June 1992 demonstrated the global dimension of these concerns and the need for harmony and cooperation.

Another reason for the growing internationalization of the environmental movement is the fact that pollution respects no borders. Effluents from smokestacks in the midwestern United States can create acid rain in eastern Canada, damaging forests and killing fish in Canadian lakes. Similarly, contaminated air blowing across Europe can damage Germany's Black Forest. Radioactive clouds from the nuclear plant accident in Chernobyl in the Ukraine strongly affected the Laplanders' reindeer herds. Nuclear waste dumped in the oceans might eventually be carried by undersea currents to the shores of many nations. The results of such cross-boundary environmental intrusions could be devastating to particular regions, often far away from the places where the troubles originated. This problem has pushed the debate between conservationist and developers to global dimensions, and international cooperative efforts are being introduced to protect the environment from these common threats.

Many governments have responded to heightened environmental concerns by developing policies and regulatory mechanisms aimed at preserving environmental quality without stifling economic expansion. The notion of sustainable development has begun to capture the attention of many leaders around the world, as they try simultaneously to respond to the demands of business and industrial leaders, who wish to continue economic expansion, and environmentalists, who increasingly see the exploitation of the earth's resources and the degradation of the ecosystem as threats to the future of themselves as well as their children and succeeding generations.

Construction for "Sustainable Development"

As the environmental movement matures, policy makers are embracing the new sustainable development paradigm for environmental management. This concept, which will be examined in more detail in chapter 2, refers to the ability to build a society whose activities meet today's needs while being designed to preserve the quality of human life for both present and future generations. The sustainable development movement has begun to affect policy and law in most environmentally related industries and should continue to have significant effects on the direction and content of environmental laws.

Sustainable development ideas will affect the construction industry in a number of important ways—providing new sources for more traditional work as well as opportunities in completely new industries. For example, construction firms can build facilities and treatment plants to convert hazardous waste to harmless materials. They might renovate existing industrial facilities requiring environmental upgrades or install more efficient energy generation systems. some companies may participate in development projects for renewable energy sources that could provide substitutes for environmentally damaging hydrocarbon fuels.

Construction firms that are willing to change their building methods or adopt new approaches that are less environmentally damaging may also find opportunities opening up for them. Builders might seek out alternate materials that preserve scarce resources and depend on more benign processing methods, or they might adopt techniques that produce less scrap. These practices may even lead to a competitive advantage for those firms that get out in front and demonstrate ways to satisfy the growing need for environmental sensitivity. Governments are starting to require less environmentally damaging methods; some industrial firms are favoring contractors using environmentally friendly methods; and consumers have been expressing a willingness to pay extra for "green" products.

While there will be myriad opportunities like this within the construction field, the major contributions of most firms are likely to be as agents of change, by helping clients in other industries to improve their own environmental performance. Chapter 6 cites a recent survey which indicates that construction firms are already feeling increasing pressures in their traditional work due to environmental requirements, and there is almost universal agreement that these will increase in the future.

The global movement toward sustainable development will tend to intensify already tough demands, inducing many industries to go well beyond just satisfying environmental regulations.

More will be expected because sustainable development represents a fundamental paradigm shift that is taking place not just in public attitudes but is reflected in the actions of both business executives and government policy makers. Industry and public agencies will be seeking ways to reduce or eliminate as much pollution and waste as possible and to find profitable uses for any remaining by-products. Since much of this activity will center on new or upgraded plants and facilities, the construction industry will play a central role in this process.

Degradation in Lesser Developed Counties (LDCs) and Eastern Europe

While initially the effects of this paradigm shift will be most apparent in the highly industrialized nations, its effects will rather quickly begin to impact the direction of development in less developed parts of the world as well, because this is where the potential will be greatest for environmental disasters. Without

INTRODUCTION

the use of advanced methods and technologies within these regions where explosive development is expected over the next few decades, there will be unprecedented increases in concentrations of pollution and waste. There are several reasons why mechanisms will be needed to apply the concepts of sustainable development to these regions.

The greatest population growth is occurring in LDCs while, at the same time, there is a massive migration to urban areas as these primarily agrarian nations try to emulate the economic success of the industrialized world. Economic expansion in these Third World nations is concentrated in metropolitan regions because cities provide convenient access to airports, highways, communications, and other necessary infrastructure that is completely lacking in rural areas. Most large factories and assembly plants are built by giant multinational corporations attracted by low-wage labor and government incentives. Then thousands of smaller enterprises spring up around these plants to provide parts, supplies, and services. Millions of workers seeking jobs are attracted to such metropolises, and the local and federal governments, unlike the more industrialized nations, do not have the resources to cope with the consequent increases in pollution and wastes. The combination of limited resources, urban sprawl, and weak governments has made many of these metropolitan areas into environmental nightmares. For example, Mexico City has grown 113 percent over the past fifteen years, and its population is expected to reach 6 million by the end of the century. Already nearly half of the world's population lives in urban areas, and it is predicted that, in the LDCs, urban populations will double over the next twenty years.

In other cases, where Third Word governments encourage impoverished people to leave overcrowded cities to carve new farmlands out of jungles, mountainsides, and other remote or undeveloped regions, the result has been the massive destruction of woodlands and rain forests. Such policies have contributed to rapid erosion with loss of topsoil in many regions, in addition to potential global warming through the elimination of vast quantities of trees and plants that help to remove carbon dioxide from the earth's atmosphere. Satellite photographs broadcast to millions of television viewers all over the globe have revealed vividly the vast devastation being wrought by land-clearing fires in the Amazon rain forests, for example.

Another cause for growing global concern is the new information coming from Eastern European and the former Soviet bloc nations emerging from the tight central dominance and information control exercised by the former USSR. Lack of adequate governmental environmental policies, coupled with the suppression of any form of opposition by indigenous populations, resulted in unprecedented levels of environmental degradation. The huge extent of these problems is only recently becoming known around the world.

Eastern Europeans are learning of the environmental progress that has already been made within the more open systems of the West, so now people in this area are increasing pressures for environmental cleanup. As efforts toward industrialization continue in LDCs with their fast-growing populations, and as expanding communications inform more and more people of environmental threats not just in their own countries but on a global scale, "green" causes will continue to gain adherents.

Even though in some cases there is only tenuous evidence of global environmental damage, along with poor understanding of the mechanisms that may be involved, governments are being forced to act because of the seriousness of the perceived threats to the planet and its ability to sustain human life. Science is only beginning to grasp the intricacies of the ecology of the earth's biosphere. For instance, the diverse chemical reactions stemming from industrial and other pollutants are far from being understood completely. Compounding the problem is the lack of good data in many areas of ecological and environmental science. But even if much more detailed and accurate date were available, global models still would have to be greatly simplified to urn efficiently on today's computers.

STRUCTURE OF THIS BOOK

We have tailored this book to help construction-industry leaders to deal more effectively with the environmental aspects of their projects. This book provides an overview and gives insights that should prove useful for entering this growing marketplace. It covers major environmental trends and key legislation that will increasingly affect all aspects of construction activity. We have incorporated presentations and discussions that took place at a symposium sponsored by Massachusetts Institute of Technology's (MIT) Center for Construction Research and Education and Japan's Hazama Corp. (see the Preface). There is additional input from numerous other sources (as cited in the References section).

We have divided the exploration of the relationship between the construction industry and the environmental movement into the following seven chapters that probe various aspects of the themes already discussed.

Chapter 1, "Environment and the Construction Industry," reveals how the construction industry has become steadily more involved in environmental markets. Some companies that specialize in areas such as sewage plants or water treatment facilities have long been part of the environmental marketplace, but many more construction companies have moved into this area in recent times. The trend started with engineering/consulting firms and has been expanding to the larger construction companies, attracted particularly by the emergence of many hazardous waste site cleanup projects, from feasibility and planning study phases to implementation. This chapter discusses strategies for participation,

INTRODUCTION

from starting up or acquiring specialized divisions or subsidiaries to collaboration and development of strategic alliances with environmental firms in the United States and abroad. We cite some of the marketplace barriers to innovative technologies, along with some potential solutions. Chapter 1 summarizes the efforts of the Environmental Protection Agency (EPA) to tackle some of these barriers, particularly in the area of communications about innovative technologies and projects. It also discusses the shift beyond end-of-pipe control solutions toward eliminating or minimizing pollution at the source.

Chapter 2, "Industrial Perspectives and Initiatives," reviews the attitudes and actions of industry as the environmental movement has taken hold and gained momentum. Examples help illustrate the remarkable shifts that are taking place as the traditional view of the environment as an infinite sink for industrial and other waste is being replaced by the emerging concepts of sustainable development. Better environmental performance is moving up on the agendas of corporate executives as fines and penalties increase and liability laws stiffen. Some companies are finding ways to turn environmental improvements into cost savings and profits. The energy arena, in particular, appears to offer great opportunities for benefits from innovative technology.

Chapter 3, "Environmental Policy and Regulations," looks at the roots of the environmental movement and summarizes some of the major legislation that has grown out of it in the United States. It also considers the various forms of regulation and shows how a shift has been taking place from a command-and-control approach to economic incentives and market-based approaches. It discusses shifting public policy, including some of the highlights of the congressionally sponsored Project 88 studies that recommended changes in the U.S. regulatory framework. There are also comments on some of the major issues addressed at the 199 Earth Summit Conference in Rio de Janeiro

Chapter 4, "Technological Challenges and Opportunities," looks at how traditional regulatory methods often inhibit the advance of environmental technologies. The actions of companies in conventional markets, where innovative R&D can allow a competitor to gain leadership in new, highly profitable market segments, are contrasted to the high risks involved for would-be innovators in a regulatory-driven marketplace. Chapter 4 describes attempts to create more incentive-based regulatory approaches to foster innovation. It also examines a wide range of environmental technologies, especially those for the important areas of solid waste and hazardous waste management and remediation. It develops the concept of the waste minimization hierarchy, including the potential for greater emphasis on waste reduction rather that treatment and disposal. There are technologies for the more efficient and cleaner generation and use of energy. The chapter describes weaknesses of the present generally adversarial relationship between government and industry, along with some efforts being made to achieve more cooperative links to foster innovative environmental science and technology.

Chapter 5, "Environmental Markets," discusses the fundamental differences between the environmental sector and other types of markets. Higher risk is involved in a regulation-driven market than in normal demand-driven markets because a shift in public policy could transform an entire market overnight, no matter how good the technology is or how much developers have invested in R&D and marketing. We break down the environmental marketplace into sectors, present market data, and discuss the dynamics driving each sector for the U.S. marketplace. Then we develop market trends for other parts of the world, including the former Soviet bloc, the Pacific Rim, and the European Economic Community.

Chapter 6, "Shifting Attitudes Create Opportunities for Construction," presents the results of a survey that was conducted by the Center for Construction Research and Education at MIT. The survey provides an inn-depth examination of the environmental market from the perspectives of construction firms and owners of constructed facilities. The chapter provides a context for analyzing the market as a whole and for looking more closely at emerging demand areas, such as hazardous waste, solid waste, and energy. The results show the type of impact and the tremendous effect of the environmental movement on the construction industry.

Chapter 7, "Corporate Strategies and Future Trends," explains that the increasing intrusion of environmental considerations into every aspect of construction work makes it vital for every construction firm to become informed about environmental regulations. At the same time, it cites some of the opportunities for construction firms that are emerging within a fast-growing environmental marketplace. While there is considerable business potential, there are also high risks and liabilities that have discouraged many firms from entering environmental markets. We present some strategic considerations for dealing with these issues.

The Appendices include the program of the MIT Symposium, summaries of the presentations, and the questionnaire used for the construction industry survey detailed in Chapter 6. A References section lists (by chapter) other sources drawn on to add meaningful material beyond that presented at the MIT Symposium.

CHAPTER 1

ENVIRONMENT AND THE CONSTRUCTION INDUSTRY

The impact of the growing environmental movement is reverberating throughout society, and the long-term effects will bring major changes to the construction industry. Many construction firms in the United States and abroad are already feeling strong pressure on their activities in traditional markets due to environmental regulations, and they expect the rules to be even tougher in the future. At the same time, however, it is becoming clear that environmental markets offer new, emerging opportunities for the construction industry. This chapter examines the growing impact of environmentalism, looking in particular at how the trend toward reduction or elimination of waste and pollution at the source will have a profound effect on the construction sector.

Construction firms have long been involved in many aspects of environmental protection. They build sewage treatment plants, water treatment systems, and waste-to-energy incinerators and install pollution controls and other waste-management equipment. The rapidly growing markets in hazardous waste remediation, waste minimization, and source control, however, are offering new challenges and opportunities, and they have recently attracted the attention of many construction companies.

Manufacturing and other industries are recognizing that eliminating waste is more than just good corporate citizenship; it can make good business sense too, because it avoids the growing costs of pollution control, waste treatment, and disposal. Policy makers and some business executives want to see a much stronger push toward sustainable development to alleviate public concerns about environmental decline, and source reduction will play a key role in striving toward this ideal. Environmental planning is increasingly being factored into new construction from the beginning, rather than tacked on as an afterthought. To accomplish this objective, contractors for both new and retrofit structures will have to work closely with their clients in industry and public agencies to help them cut back on waste and pollution and to provide treatment for any waste generation that is unavoidable. These shifts in attitudes and practices are likely to become more acute as construction firms enter into newly expanded and rapidly growing environmental markets. These markets should offer new business opportunities to construction firms that gain the specialized knowledge needed to cope with these problems.

Construction's role in helping to clean up the environment is rapidly expanding. Industrial firms, struggling to cope with a steadily increasing burden of environmental regulation, are beginning to shift away from the traditional waste-management approach toward pollution prevention and source control. Construction companies will increasingly be called upon to help industry make the transition. The results will be far-reaching. While entry into environmental markets has not been easy, it has generally had little impact on the core functions in a construction organization. By contrast, moving toward source control is likely to have significant effects on almost every phase of the business. A closer look at the evolution of the dynamic environmental marketplace will help reveal the changing role of the construction industry and the impact that can be expected on company structures and relationships.

So far construction firms have been able to penetrate the environmental field by adding a few skills and some new knowledge, often within a specialized division or subsidiary. Such groups combine traditional construction lore with newer methods applicable to particular aspects of the market. These affiliates may also help keep the parent firm posted about additional business opportunities, perhaps due to potential regulatory shifts. Fast growth and high profitability in several segments of the environmental marketplace provide powerful lures to an industry that has seen once-booming commercial and industrial construction virtually dry up in recessionary times. The up-tempo pace in the environmental sector also promises to continue for years and even decades to come.

Progress in market development and expansion will depend on innovative technology and processes, but several barriers built into the current structure of the environmental industry and its corresponding contracting practices are hindering innovation. A good example is the dynamic hazardous waste disposal market. Existing waste-management and waste-hauling firms saw an opportunity to expand their traditional businesses into this fast-growth sector. But they found this emerging market to be far more demanding. In the early days, the services needed were mainly hazardous waste handling, transportation, and "disposal" which consisted of landfilling, deep-well injection, or incineration. As experience has been gained in cleaning up hazardous waste sites, however, requirements have steadily stiffened. Specialized technologies are now required to cost-effectively destroy, detoxify, or encapsulate heavy metals, toxic organics, or other hazardous substances. Cleanups often involve large quantities of tainted soil, created by leaking barrels or other buried containers, that must be decontaminated. Then, after treatment has been completed, conclusive testing methods are essential to verify that all toxic materials have been rendered harmless.

Even after acquiring the necessary skills, these fledgling hazardous waste firms found that they faced long bureaucratic delays before work could begin. Labyrinthine regulations required extensive planning and feasibility studies, leading to a choice of a cleanup method acceptable to the EPA, which was then incorporated in a Record of Decision (ROD). In addition, remediation entailed extensive liabilities calling for costly bonding and insurance coverage. This combination of specialized skills, delays, and high costs discouraged some waste-management firms and even caused a number of them to fail. In 1981 a Frost & Sullivan survey found that seven companies did 40–60 percent of all the hazardous waste-management work, but by 1992 only two of these (International Technology Corp. and Chemical Waste Management, Inc.) remained in business.

As money was spent on the cleanup effort, other types of firms entered the market. Because of the emphasis on feasibility studies and planning, consulting/engineering and geotechnical companies were the first outsiders to be attracted to the business. In more recent times, as a growing number of National Priority List hazardous waste sites move from the planning stage into actual cleanup operations, construction companies have begun to enter the market. Some larger construction firms offer advantages over smaller waste-handling companies. They offer financial strength and experience in managing large-scale, long-duration projects. Opportunities are increasing for construction firms as the cleanup effort moves into full swing. While some $10 billion had been spent on remediation work through late 1991, only slightly more than 60 sites had been completely cleaned up. Now that pace is quickening. The EPA's goals are to have 140 sites cleaned up by the end of 1992, rising to 200 by the end of 1993, and reaching a minimum of 600 by the turn of the century.

Skills and knowledge gained by U.S. construction firms in the environmental area may prove valuable in overseas markets as the "green" movement spreads around the globe. But operating overseas will require construction firms to adapt to the new international business climate based on interlinked telecommunications, information, and decision systems. Gaining access to global markets also may require partnerships and alliances with foreign firms as world markets move towards semiclosed trading blocs.

CONSTRUCTION FIRMS NEED TO CHANGE THEIR STRUCTURE

Environmental markets, especially with the new trend toward source control and its added complexities, may call for construction organizations to go well beyond setting up separate environmental divisions and vying for specific contracts. It may lead to changes in the fundamental structure of the industry itself as it gears up to meet the challenges of achieving the so-far elusive goal of "sustainable development," which will be explored in more detail in Chapter 2. Environmental considerations will become a critical portion of most major industrial and commercial development projects in the future, and construction companies will have to be prepared to meet the new demands this will entail. If any kind of waste or pollutants are involved, there is almost certain to be demands for fresh approaches that reduce their generation or that treat any that are produced right at the source. Rules for handling and disposing of waste and other hazardous materials have become very tough. There is greater appreciation of the true costs of dealing with hazardous waste, and the industry is eager to find ways to reduce or eliminate it from operations whenever possible.

While most work done by construction companies follows pre-engineered and architectural plans, a construction contractor will still be responsible for ensuring that environmental requirements are properly considered. While the primary impact of the environmental movement so far has been predominantly on engineering/design firms and on specialized firms dealing with environmental aspects of a project, the shift toward elimination or minimization of pollutants and waste at the source will require greater collaboration between these different segments of the total construction industry. Construction firms will have to be prepared to make useful contributions to their clients from the beginning of a new project right through every stage. This will require environmental skills and knowledge to become embedded into every department of a construction firm, rather than simply being relegated to a specialized division, subsidiary, or group. Trends in the hazardous waste arena alone illustrate the need for broadening the scope of environmental capabilities across construction organizations.

Most estimates place the amount of hazardous waste generated in the United States each year at 250–500 million tons, or between 1 and 2 tons for every U.S. citizen! Somewhat over three-quarters of this amount is generated by three industry groups: chemicals and allied products, primary metals, and petroleum and coal products—all traditionally major clients of the construction industry. These industries are already becoming quite active in source control efforts. Under the EPA Industrial Toxics Program, or "33–50 Program," some 600 manufacturers were asked to reduce pollution voluntarily from 1991 levels of 17 toxic chemicals by 33 percent by 1992 and by 50 percent by 1995. The Chemical Manufacturers Association has developed a Responsible Care Program that lists pollution prevention as one of the guiding principles for its 185 members. Source control is also integral to a voluntary corporate ethics code drafted by the Coalition of Environmentally Responsible Economies (CERES), which covers a wide range of issues, from the wise use of natural resources to agreements on compensation for any damages. Although source control has not yet reached the federal level, several states have instituted strong waste and pollution prevention legislation. Massachusetts, for example, under the state's Toxic Use Reduction Act of 1989, has a mandated goal of a 50-percent reduction in the use of toxics by 1997. William Reilly, EPA administrator during the Bush administration, suggested that, through source control, "companies can save on waste management, reduce the use of raw materials, and minimize liability." He also pointed out that reducing pollution, emissions, and other waste will relieve firms of regulatory burdens and yield other benefits that could help make them more competitive in world markets. Source control has forced manufacturing and other industries to seek the cooperation of the construction sector in all aspects of project development. In other words, demand for better project planning, for utilization and installation of new technologies, and for development of new processes that either eliminate or substantially reduce waste will compel the construction industry to integrate environmental concerns at every stage of project development.

The environmental movement, however, is only one of the forces now reshaping the construction industry. Its influence must be viewed within the context of many interrelated changes. Business has been globalized, and advances in telecommunications and information technology provide worldwide networks that are having far-reaching influence in many economic sectors, including capital and finance, the procurement of materials and services, and the monitoring of socioeconomic developments in virtually every nation. Construction firms are an integral part of this emerging global business framework. They can look for opportunities all over the globe to obtain materials, equipment, and services. Even low-cost labor is increasingly flowing from regions with excess population to labor-short areas such as the Middle East, Japan, and some parts of Western Europe.

International competitors are beginning to assemble the needed capital, goods, services, and labor to vie for projects in any of the four major construction markets: Western Europe, Eastern Europe (including the nations of the former USSR), Japan, and the United States. This global competition may be constrained somewhat by trading blocs that now appear to be taking shape, starting with the European Economic Community and potentially expanding to include the Far East, North America, and possibly the remnants of the old Soviet bloc. Tapping each of these markets may require new alliances and partnerships, which are already being forged.

Information networks are altering the ways that construction firms operate within this new global economic setting. Telecomputing systems allow closer monitoring of far-flung projects with the ability to gain a quicker, more comprehensive overview of work status and potential problems. They also give management more opportunities to develop comprehensive strategic plans incorporating economic and market conditions, competitive factors, widely scattered suppliers, and new technologies. Firms are better able to monitor their progress and adjust strategies accordingly. Links with financial institutions are also becoming a vital part of the business. A construction contractor may be able to win business by putting together a complete package for a client, including financing and even including mechanisms for dealing with fluctuations in interest and currency exchange rates.

There are other ways in which the construction industry is changing because of the growing dependence on information systems linked by telecommunications networks. Specialized knowledge has become a more critical competitive factor as construction has grown more sophisticated and knowledge-intensive. In the past, much of this proprietary edge was provided by knowledgeable individuals who were free to move from firm to firm. More and more, however, knowledge is being assembled within organizational structures and interlinked telecomputing systems. Organizations are strengthening their competitiveness through intelligent data gathering, large computerized databases, and development of specialized software. These assets can be made proprietary and subject to licensing and patent rights.

Global concerns over the environment will offer tremendous market opportunities but, to take best advantage of them, construction firms will have to become adept at operating within this new business framework. Penetrating growing segments of the environmental market may require a construction company to develop specialized competencies in tasks such as treating contaminated soil in situ, building leak-proof landfills, or devising better ways to eliminate or neutralize pollutants. But along with expanded opportunities in such specialized areas, there will also be greatly increased local and regional scrutiny of virtually all construction work.

While there has been a recent decline in the number of megaprojects as real estate has been hard hit by a global recession, those projects that are being planned are bigger in scope ($5–10 billion each) and are in or near urban areas. Developments like these will be subjected to even more stringent environmental requirements than in the past. It will not be enough for a construction firm to select a portion of the environmental market and then concentrate all its efforts in that area. An awareness of the steadily widening scope of environmental rules will be essential, not just in a firm's own domestic market but worldwide. A quick review of the changes that have been brought about by the environmental movement will help in exploring the impact the construction industry might expect in the future.

TIGHTER REGULATIONS SHIFT THE ONUS TO CONSTRUCTION CONTRACTORS

In past decades, often under the duress of tightening government regulations, pollution controls or waste-management systems were added on as afterthoughts well after facilities were built. The popular view among executives was that environmental controls were external to the main business, diverting capital investment away from productive uses and into facilities that added nothing of value to a firm's output of goods or services. Loose enforcement of environmental laws sometimes led to unchecked abuses. If waste generators felt they could get away with avoiding regulatory strictures, many times they resisted making environmental investments until well after corrective action was mandated. Practices such as "midnight dumping" and the discharge of untreated toxics into streams and rivers became widespread. Eventually the cumulative effect of this failure to protect the environment adequately led to the public outcry about filthy waterways, smog-laden air, and acid rain. As public anger mounted, the harsh spotlight of publicity occasionally led to public embarrassment and lost business for companies perceived to be ecologically irresponsible.

Health hazards stemming from past practices such as these have now revealed themselves and been recognized. Cleanups have been mandated for potential problem sites, such as dumps where hazardous fluids are buried in containers that can corrode and leak; holding ponds that contain heavy metals and other toxic mining wastes that are washed into mountain streams by the spring melt; and buildings containing asbestos insulation,. To avoid any repetition of such environmental abuses in the future, rules have been tightened, the scope of environmental regulation broadened, and enforcement stepped up. On Long Island in New York, for example, authorities conducted a five-month sting operation ending early in 1992 that resulted in arrests and indictments of four companies and six individuals who had paid minimal fees to have barrels of toxic waste dumped illegally. In one incident an electroplating company paid only $450 for illicit disposal of eight drums of toxic cyanide, a by-product of some plating processes. Proper disposal would have cost over $10,000.

Not only have fines for such actions risen steeply, but numerous executives responsible for activities that were judged to have endangered public health or safety have gone to jail in recent years. During the 1970s, only 25 criminal environmental cases were prosecuted at the federal level. By contrast, much tougher enforcement has led to 703 federal trials over the period from 1983 to 1990, with about half of those cases dealing with hazardous waste violations. Since 1983 some cumulative 348 years of prison terms have been meted out for environmental crimes.

Stepped-up enforcement by the EPA has also led to increased private funding of the cleanup of Superfund hazardous waste sites. In fiscal year 1991 a record $5.1 billion was spent in private-party funding. Some 63 percent of Superfund cleanup costs are now covered by funds from private parties compared to only 37 percent in 1987. Contributions increased dramatically after the EPA began a new policy of "enforcement first" in 1989, as opposed to the EPA doing cleanups on its own and then trying to collect afterwards. The size of settlements has been escalating, as can be seen from an agreement announced by the EPA in December 1991, which at the time was the largest ever for a single Superfund site cleanup. The settlement stands at more than $200 million from 170 parties for cleanup of an Operating Industries, Inc., site in Monterey Park, California, along with the reimbursement of all costs incurred by state and federal agencies.

With environmental rules and their enforcement becoming steadily more pervasive, the attitudes of the customers of construction firms, whether in industry, government, or commercial development, have been changing. Clients now expect construction companies to be knowledgeable about applicable environmental rules or permits, even though pollution control or waste management may not actually be designated as a part of a project. Construction contractors cannot assume that lawyers, consultants, or other specialists have addressed all pertinent environmental issues. If the contractor fails to deal properly with applicable regulations, even though a client was unaware of them, the consequences may be delays, increased costs, or even cancellation of the job.

As environmental consciousness rises, customers for construction work expect contractors to be knowledgeable and to act in a way that extends beyond simply following regulations. Some clients may also wish to address other ecological concerns, such as energy efficiency and the recycling of solvents, process by-products, or the sprue from die-cut materials. Advanced structural engineering methods and the use of lighter-weight but stronger alternatives have helped bring a marked reduction in the amounts of material used per unit of gross national product (GNP) in industrialized societies. Clients expect to be able to take advantage of such advances by hiring construction contractors that remain aware of the latest materials and building technologies.

THE PUSH IS ON TO ELIMINATE WASTE AT THE SOURCE

Above all, industrial firms want to find ways to eliminate waste. Many manufacturers are hoping to install what are being called "clean production technologies" in their plants. If some pollutants are unavoidable, they want to avoid facing long-term liabilities due to inadequate treatment or disposal. There is a growing acceptance of a hierarchy of approaches to waste management, starting with source control if it is feasible. If not, the next preference is for recycling, then treatment, and finally, as a last resort, disposal. There are five basic approaches to controlling waste at the source:

1. Change raw materials used in production.
2. Change production technology and equipment.
3. Improve production operations and procedures.
4. Recycle waste within the plant.
5. Redesign or reformulate end products.

Construction companies handling such jobs, whether building new plants or retrofitting and retooling existing facilities, need to work with clients to come up with processes that are as pollution-free as possible. They will sometimes need to customize designs to a particular manufacturing process. In other instances, a construction firm may need a completely innovative approach, as was the case for an AT&T plant that will be discussed in the next chapter. [The contractor must also be knowledgeable about treating and disposing any residual waste and pollution.]

This trend toward source control is not just being driven by altruism, although there are a few industrial firms that have decided that helping to keep the environment clean is part of responsible corporate citizenship. An even more powerful incentive in the push toward what is sometimes called "industrial ecology" is the realization that environmental improvements can also make good business sense. It can be a lot cheaper to build plants and design production facilities so that waste and pollution are eliminated in the first place than to have to pay the high costs of controlling, treating, transporting, and properly disposing of waste later. Even beyond these direct costs is the specter of assigning long-term liabilities to the originator of any waste that prove harmful at any time in the future. Companies that develop a reputation for environmental responsibility also find that they frequently get a more friendly reception from potential customers.

Sometimes it is possible to alter processes so that substances that had been considered "waste" can instead be turned into profitable by-products. In many power plants in Europe, for example, processes used for flue-gas desulfurization in many power plants yield calcium sulfate, or gypsum. Most of this gypsum is sold for use in wallboards or as a supplement to cement. Several European countries along with Japan have achieved very high utilization of desulfurization by-products in construction. So far, however, there has been little effort to do this in the United States, even though systems that yield useful construction materials from flue-gas desulfurization are available. One system developed by Japan's Shoida is now being marketed in this country by Bechtel, for example.

Moving toward more effective environmental source control changes many things about the way an industrial firm, government agency, or commercial developer does business. To address these issues from the start of any project requires much greater involvement with those in the construction arena. Adjustments may be needed at every stage, from architectural and site planning through project management, engineering, and construction, and then on into operations and maintenance. To serve such clients well, construction firms need to add environmental considerations to an already wide range of skills. Beyond civil engineering, contractors may need to add capabilities in areas such as hydrology, air quality, biotechnology, ecology, bacteriology, and health and safety. For industrial projects, the mix of talents required may include chemical engineering, industrial engineering, materials handling, and in many cases detailed knowledge of manufacturing and process engineering. Contractors also must keep abreast of the state of the art in pertinent areas of environmental science and engineering.

Even with these added skills, however, source control cannot be achieved without increased interaction between construction firms and their customers. Help from specialists within the client's organization is essential if waste and pollution are to be reduced or eliminated without impeding productivity. Instead of acting as passive clients, plant owners and operators need to share information and contribute to effective solutions.

TEAMWORK APPROACH

This collaborative approach to mutual problem solving is extending from the customer interface right across the construction industry itself. Linkages between firms is becoming common as clients frequently seek contractors that can handle every phase of a project. Consulting/engineering firms are increasingly forming partnerships with construction companies as well as with specialists capable of handling different facets of environmental jobs. Even in public works, where there are often legal or institutional barriers to such alliances, more agencies and municipalities are recognizing the potential advantages and, as a result restrictions are being relaxed or eliminated. Under the Superfund program, for example, historically the EPA's policy was to separate a project into sequential phases, such as studies, design, and construction, with different companies being responsible for each phase. Detailed specifications would be developed for a project without the planners even knowing which contractors would perform the actual work. Results were often poor. Therefore, recently the Superfund program began to allow single firms to handle all phases throughout the entire course of a project. There is also more stress on smoothly integrating the work of design and construction. If those doing the planning, specifications, and design can work as a team with those who will handle the actual construction and installation, many loose ends and the seemingly endless delays and backtracking needed to resolve conflicts can be eliminated before work even begins. If a designer knows the contractor assigned to do the work, the designer can tailor the project to that organization's strengths or special capabilities. When construction contractors get into a project early, they can often make suggestions that keep costs down and eliminate impractical or unconstructable design elements. Any design requires many trade-offs and compromises, and by bringing all parties to the table early, including designers, engineers, builders, and the client, these issues can be resolved effectively before the work begins. This can speed up projects and lower costs and greatly improve communications.

By forming partnerships, consulting/engineering and construction firms can also offer a single interface to the client, with a single, consolidated fee for an overall project rather than piecemeal pricing. Sometimes, when special skills are needed, even firms that would otherwise be considered competitors are brought into a project team to handle specific phases of the work. This trend toward alliances and team-based project management allows more flexibility because of the wider mix of skills available; much better communications; and faster, more flexible response to clients' needs.

Firms in the construction business have to adjust to this new mode of operation. They must go beyond strictly technical issues, developing the ability to manage, motivate, and communicate with those possessing complementary skills. A project may involve legal issues, environmental impact studies, labor relations, and technical matters that require effective interaction with specialists with vastly different backgrounds and training.

There is another factor that has increased competition in the environmental marketplace—this marketplace has continued to grow as other markets have been hit hard by the recession. A proliferation of smaller start-ups joined the consulting/engineering and geotechnical firms that branched out into environmental work during the early buildup of the business. Now, as larger cleanup jobs are moving toward implementation, even a few defense contractors are beginning to compete with the large construction firms now moving into the remediation market. Military spending is declining, so these defense firms are looking for other opportunities for their project management expertise. While the entry of industry giants does makes it tougher to compete, especially for the smaller firms, there is still a need for niche players, so the small firms can survive. Many clients even prefer to work with a local, dedicated operation, whether a field office of a larger company or a small firm in the area where the work must be done.

Engineering/consulting firms and others that have specialized in environmental work are finding that they must change their approaches in order to win against the much tougher competition they are facing for new projects. With experience in planning cleanups accumulating, there is not as much need for developing unique approaches for the remediation of each site. Former EPA administrator Reilly believes that the Superfund program had reached a point where "some standardized procedures and guidance can be developed." Fees also are coming down, not only because of the ability to draw on past experience but also due to the increased competition in the marketplace.

Another factor that existing environmental contractors must contend with is that more clients are looking for opportunities to do "one-stop shopping." Environmental firms may find that they need to form alliances with construction companies and other specialty firms in order to offer the end-to-end capabilities now being sought. Links to successful construction firms can also help engineering/consulting and other environmental specialists to position themselves for the trend toward source control. These construction companies, particularly those that build process plants or refineries, can provide entry to this emerging market segment. This trend toward alliances is sometimes bringing together U.S. consulting/engineering firms and international construction companies.

International firms are interested in forming partnerships with American firms because the United States has become more advanced in methods and technologies since it has addressed serious environmental problems earlier than most other countries. This is particularly true for hazardous waste management. Foreign firms want to gain knowledge and expertise from experienced U.S. companies so that they can then put these skills to use in their own markets or in other foreign countries. It is interesting to note that in some cases the technology used by the U.S. partners is developed in Europe. However, the U.S. firms have the expertise needed to use the technology and to effectively manage the construction site.

Why should construction companies want to forge links to environmental firms? We can find the answer by looking at the strengths of each of these types of firms and noting the changing market demand that will require a combination of these assets.

With their broad range of capabilities, many construction companies are already structured to handle multiple phases of a project, from feasibility studies and architectural work to engineering and implementation. Their capabilities in project management in particular are far superior to those of most environmental firms—these include broad experience in project scheduling, estimating, project supervision, contract preparation, and labor relations. So far, the average cost for cleaning up a National Priority List site has been in the $20–30 million range, making these specialized skills not quite so critical, but costs and complexity are expected to increase. A dramatic example of the predicted escalation is the settlement reached for cleaning up the Rocky Mountain Arsenal. This project will probably cost more than $1 billion, with Shell Oil and the U.S. Army paying for the cleanup. Cleanup megaprojects such as this one will require the kinds of project management skills offered by construction companies. For smaller projects, such as cleaning up a leaking underground tank or even an abandoned landfill, there is little need for sophisticated project management techniques, but they do offer excellent opportunities for local firms.

Why Alliances Make Sense in the Environmental Marketplace

In spite of their strengths for handling larger projects, even major construction firms lack some of the capabilities offered by engineering firms and others that have specialized in environmental work. They will still have to assemble a wide variety of services required to investigate and develop a cleanup plan. Preparation to move into full participation in the remediation market calls for adding these complementary capabilities. Environmental firms, for example, may offer specialized services such as soil sampling along with soil and groundwater analysis, geotechnical and engineering design services, waste transportation, well drilling, remediation technologies, and off-site treatment and disposal services. A construction firm that assembled an array of such services in-house would have to cope with a lengthy learning period and the costly process of establishing a track record. Forming an alliance with an existing environmental firm offers quicker access to the marketplace. A long-term commitment to the environmental market may even suggest one or more acquisitions to bring the needed skills into a large construction firm. Operating in this arena also calls for specialized knowledge in areas such as risk management, environmental law, public relations, and regulatory legislation and enforcement, all of which construction firms may readily acquire through strategic alliances.

Even though construction firms are experienced in handling risk for large buildings, bridges, tunnels, and the like, the nature of potential liabilities are quite different in the environmental field. Effects on workers from harmful chemicals, such as higher rates of cancer or birth defects, may not occur until many years or even decades after disposal. A possible explosion in a tunnel or a bridge collapse is a tangible event with somewhat predictable damages. Long-term health effects are much more difficult to quantify, or even identify, with almost unlimited potential damages. Setting up separate entities to operate in the environmental field is not proving to be a viable way to shield the parent firm from potentially devastating liabilities. Also, courts seldom accept attempts to reassign liabilities to suppliers or subcontractors. The trend is toward assigning the major damages to any organization with deep pockets, no matter how incidental its involvement.

With such serious potential liabilities, along with the intricacies of environmental regulations, it's vital that decision makers in any larger firm operating in this field have access to competent legal counsel. Some firms in the industry depend on legal advice not just to comply with applicable regulations but to avoid any question of carelessness or endangerment of health and safety. A few companies, however, have used legal departments to fend off challenges to their operations; they have found it more profitable to cut corners, keeping costs down while paying any fines incurred as a result of their activities. Since fines and penalties are increasing and since executives are being held personally responsible for the actions of their organizations, this strategy is becoming increasingly risky. As public attitudes, along with those of potential customers, are becoming more critical of environmental abusers, it will prove more difficult to continue doing business in this fashion. Using legal assistance to avoid such problems in the first place may prove to be a far less costly strategy. Many large environmental firms have recognized this and have established competent and well-staffed legal departments.

Public relations can also be an important element for s... mental arena. It can be vital not just to counter negative... ensure that positive actions to preserve or improve enviro... get the attention of the media. Companies operating in thi... stay abreast of regulatory changes. Firms with far-flung op... posted on local, state, and federal regulatory requirements. ...ment representatives, while performing lobbying functions, can also keep their organizations aware of regulatory trends that could offer new business opportunities.

Technology is often a critical element for success in the growing remediation market as well as other environmental sectors. In the remediation area, some 210 different technologies have been specified in EPA Records of Decisions. These can be roughly divided into five categories: thermal treatment, solidification/stabilization, physical separation, chemical treatment, and biodegradation. As environmental markets become more competitive, it is essential for a company hoping to expand in this area to remain up to date on the latest techniques, including those already in use and those emerging from research laboratories around the world. It might make sense to add a specialized technology company or to form alliances between environmental and construction companies in order to add strength in pertinent technologies. Thus, both environmental firms and construction companies interested in entering or expanding in this market can find mutual benefit in forming some type of partnership.

Another approach to adding strength and expertise is to make an acquisition. This is a faster and more convenient route than attempting to develop the necessary competencies internally, and is the most common method of forming alliances. The acquiring construction company cannot step right in and take over the activities of the newly acquired subsidiary because it does not have sufficient experience in the environmental field. Instead it must set up communications links that will help increase its understanding of the business and provide opportunities for the two cultures to develop a common ground. It will also help to identify mechanisms for the eventual integration of all activities. Meanwhile, the parent company should allow the subsidiary to operate fairly independently. If the corporate cultures of the two firms are too disparate, friction is inevitable and valuable talent may be lost. Thus, it is important to try to determine whether there will be a good fit before making the acquisition.

An alternative to making an acquisition is hiring specialists with the required skills and building up an internal competence in selected environmental sectors. This will require higher salaries to lure good people away from other organizations; it will take longer than entry through acquisition; and there may be some false starts before a cohesive, efficient organization is in place.

There is another approach that avoids the problems of either making an acquisition or building a new environmental unit—that is to form strategic alliances with other firms. Joint ventures of this type are becoming more common for larger projects. This is a way to share risks and costs without incurring the long-term obligations of a full-fledged subsidiary. There is always the risk of a partner not performing as agreed upon. In some cases a partner may be using the project as an opportunity to become a competitor in the market by gaining complementary skills and even hiring away valuable employees. The chances of such pitfalls can be minimized by carefully screening potential partners and signing contracts that serve to protect both parties from unfair practices.

NUMEROUS BARRIERS STIFLE INNOVATION

Attaining the so-far elusive goal of "sustainable development" will require a steadily advancing set of technologies. These will be needed not only for controlling and treating any pollution or waste generated, but also for reducing or eliminating the waste in the first place. Yet those involved in the field, both environmental companies and owners of facilities that must deal with pollutants or waste, complain that our present system of regulation and enforcement actually discourages innovation. The barriers and difficulties faced by engineering and construction firms hoping to innovate in this field are also an intertwined combination of technical, financial, regulatory, and institutional/legal factors.

On the technical side, in the hazardous waste area the unknowns are enormous. Contamination is normally underground, limiting the available methods for data collection and modeling. Even with the best analytical techniques, risk assessments are uncertain. Questions also remain about the effectiveness of today's remedial techniques.

Adequate modeling of seepage and contamination presents a serious problem in trying to devise preventive strategies because, even with the most advanced methods for subsurface soil sampling and hydrogeology, engineers still cannot obtain sufficient data. Yet they might be asked to predict, for example, the flow direction of an underground chemical plume, indicating where it could be extracted. Finding adequate methods for such microscale problems may take huge investments and years of research. In spite of this limited knowledge, engineering firms are asked to come up with risk assessments that define and establish standards for acceptable cleanup levels. They are forced to use guesswork, and, if estimates are too conservative, they may be specifying cleanup technologies that don't even exist yet.

Even if an engineering company sees opportunities for developing innovative technology, what about the financing? Management is being asked to fund developments in a highly charged public arena swirling with issues such as untested liabilities; the potential for complex, costly litigation; and the need for surety bonds. If research funds are not available internally, firms find outside investors to be extremely wary of the environmental industry with all its uncertainties. Outside investors have seen how public opposition has squashed plans for potentially highly profitable waste-to-energy incinerators, for example. Snags and delays that forestall or eliminate the potential payback from investments in this market have been so common that few sources of outside capital remain.

Institutional barriers also discourage innovators. The complexity and fluidity of the regulatory framework can seriously handicap efforts to push new developments forward. In the hazardous waste area, for example, the Comprehensive Environmental Response, Compensation and Liability Act of 1980 (CERCLA or the Superfund Act) and the Resource Conservation and Recovery Act of 1976 (RCRA) each has its own requirements for risk assessments. When a vendor must operate under both programs, different agencies may come up with varying interpretations of the rules.

On top of problems such as these, individual state laws and regulations can be even tougher than federal requirements. Technologies demonstrated to be effective in one state or EPA region may not be accepted in another. Engineering firms complain that the EPA doesn't have any formal program to ensure that current guidelines are available on all aspects of hazardous waste management and site remediation. Groundwater-monitoring issues were covered or touched on in no less than fifty-eight guidance manuals and twenty memorandums, for instance, until the EPA finally came up with a Technical Enforcement Guidance Document.

Even carefully following the rules may not be enough to prevent liabilities for firms engaged in cleanups. One problem is that what constitutes "normal and accepted practices" is not clearly defined, opening the way to negligence suits. It isn't enough to prove that every reasonable precaution has been taken. Courts have found that persons who engage in "hazardous" activities are still open to strict liability suits. Also, because of the doctrine of "joint and several liability," cleanup firms that participate in an investigation or design effort may be held liable along with the original contaminator if they fail to identify all the contamination involved or if the selected remedy fails to work.

What's Needed to Spur Innovation?

Even if such barriers were eliminated, the plodding pace of innovation would not likely pick up very much. In past eras, innovation often came in the form of ingenious inventions satisfying widely recognized needs. In more recent times, with the growing complexity of our mechanized, computerized world, planned innovation has become more common. By training varied specialists in many narrow technical areas and then letting them work on their own segments within larger programs, innovation can proceed without the need for an overall guiding genius. This sort of planned innovation is how the technology was developed for landing on the moon, for example.

Achieving such advances requires a good deal of interaction among researchers delving into different aspects of multifaceted problems, such as those encountered in the environmental arena. This is especially vital for moving scientific discoveries out into the engineering world so that they can be applied to practical designs. Unfortunately, the organizational structures and philosophies of the types of engineering companies involved in the environmental industry do not encourage this sort of interaction. Often different facets of large projects are handled by separate companies. Even within a diversified firm, tasks are usually isolated so that separate divisions or subsidiaries work on them, often with minimal coordination and interaction.

Industy has failed to recognize that an invention is an idea while an innovation is a process, according to Howard B. Stussman, editor-in-chief of *Engineering News Record*. He points out that few can name the inventor of the automobile: Nicholas Joseph Cugnot, a Frenchman who came up with his invention in 1769. But everyone knows about Henry Ford, whose method of mass production finally turned the invention into a major industry. Stussman suggests that the advances needed for meeting the challenges of cleaning up the environment will likely come from organizations that practice what management guru Peter Drucker calls "the Dos of innovation."

1. Purposeful innovation begins with the analysis of opportunities, an organized search done on a regular, systematic basis.
2. Innovation is both conceptual and perceptual, drawing on both the left and right sides of the brain. Successful innovators and innovative companies look at both numbers and people. They encourage innovators by offering challenge, responsibility, resources, sponsorship, and rewards.
3. Effective innovations are simple and focused.
4. Effective innovations start small, requiring little money and only a few people, beginning with only a small or limited market. Otherwise, there is not enough time to make the adjustments and changes almost always needed for innovations to succeed.
5. Successful innovation aims at leadership. If it does not aim high enough at the beginning, it is unlikely to be innovative enough to establish itself.

Such organizations are rare in American industry, and Stussman points out that a macro view of the construction and environmental fields reveals that the expertise essential for innovative advances is distributed across a wide range of diverse organizations. Achieving the nation's environmental goals, he believes, will require a national innovation policy. The important elements of such a policy, according to Stussman, would be:

1. up-front R&D funding for innovative ideas and projects
2. links among the diverse organizations needed to achieve innovation, including universities, nonprofit organizations, government laboratories and agencies, and industry
3. incentives for marketing innovations

Frank Press, president of the National Academy of Sciences, shares similar views concerning the appraisal of the fragmented nature of the knowledge required for innovation. He suggests that, in the environmental arena as well as other areas of advanced technology, the government might be able to build and sustain an organizational complex and also facilitate the formulation of innovation-specific networks. Additionally, according to Press, government could do such things as: assist in the underwriting of risks associated with process and product innovation by providing long-term, low-interest funding; help identify and develop markets; provide loan money for the foreign purchase of U.S. products and make such trade a central concern of foreign policy.

Another factor working against innovation, according to some in the environmental field, is that work is generally given to the lowest bidders. Since R&D is costly, companies that want to take innovative approaches must make large investments on their own or they are unlikely to win environmental jobs. Contracts also are not structured to protect a company that makes the extra effort to try an innovative approach. If useful advances do result, they are likely to be picked up by other contractors that have not made the investment in R&D and experimentation. Larger companies can ensure that they protect such work with patents, and then can offer their technology for licensing fees. But smaller firms with limited resources for innovation may find that, before they can get to market, larger companies are able to take the kernel of new ideas, develop them more quickly, and market them more effectively. Without some form of contractual protection, this risk of lost opportunities can be particularly troublesome in larger, multifaceted projects in which numerous contractors must share information. Before making heavy commitments to R&D, investors need to be sure that they will be able to get adequate payback for invested funds.

Government agencies are beginning to address some of these institutional barriers, according to Walter W. Kovalick, Jr., director of the Technology Innovation Office of the EPA. Projects involving public/private collaboration are now being set up to demonstrate new technologies on government sites with industrial sponsorship under the EPA's Superfund Innovative Technology Evaluation (SITE) program. Federal facilities agreements supervised by the EPA are substituted for the usual permits. Although this program does not fund initial stages of technology development, it does help fund demonstrations through a cost-sharing process. Developers pay for mobilization and operation of the technology demonstrations while the EPA pays for planning, sampling and analysis, quality assurance, and report preparation. SITE projects give developers an opportunity to introduce new and emerging technologies in the environmental market while also producing information that can be used nationally, both by other engineers and industrial users and by policy makers developing rules and regulations governing cleanup operations. More than seventy technologies are enrolled in the program. Also, there are more than thirty participants in the SITE Emerging Technologies Program, which provides funds up to $300,000 over a two-year period for bench or pilot development of advanced technologies.

Also, in mid-1991, the EPA issued a directive on innovative technologies containing seven initiatives. The directive urges EPA regions to seek cost-effective solutions, but with recognition that first-time applications of innovative technologies might fail to perform as desired so that reengineering and redesign are likely to be required. It points out that the managers and coordinators involved in innovative projects must be encouraged to view their work as "career opportunities" rather than as "career risks." This suggests that at the same time that the EPA is striving to change management attitudes toward risk taking, industry needs to ensure that consulting and construction firms will be given similar encouragement. One opportunity, Kovalick suggests, lies in thousands of private industrial sites that corporations will have to clean up on their own. This will give companies opportunities to try out innovative technologies that then may prove marketable.

In addition, the Department of Energy (DOE) provides funds to developers of new technologies to assist in adapting them for use in remediation of DOE sites, and it sponsors major "Integrated Demonstrations" at a number of its laboratories.

The military has many thousands of hazardous waste sites that match the problems faced at corporate locations. At a recent meeting of the EPA, the armed services, and thirty of the largest U.S. firms, it was agreed that user groups could be formed to exchange information about innovative approaches to similar cleanups. Military participants expressed willingness to work with industry on new methods for attacking such problems, with the involvement of the EPA if desired. The Federal Technology Transfer Act also provides an opportunity outside the normal contracting process for private industry and government laboratories to work together on joint projects.

BETTER COMMUNICATIONS ON NEW TECHNOLOGIES NEEDED

The EPA now recognizes the lack of adequate communications about innovative methods as another major barrier that must be addressed. All of the cleanup work has been done with some 200 firms, with many more companies wishing to participate in the program. But the EPA finds that mechanisms for exchanging useful information about the latest methods are sadly lacking. In surveying some of the largest contractors with nationwide facilities, the EPA discovered arcane methods for providing such data to engineers who might make use of it. In one firm, one individual in the organization was assigned to attend conferences, scan the literature, and in general keep posted on the latest developments. But newer engineers in the firm might not even be aware of this "source." In another firm, a small group was assigned to scan magazines and other literature, find reports on innovations, and then enter abstracts into a computer system. Unfortunately, the firm had not yet devised very good methods for finding and retrieving the proper information when it might be useful.

To address the shortcomings in communications, the EPA is working on a variety of approaches to make information on innovative environmental technology easily and widely accessible. Twice a year the agency puts out a bibliography of all the publications it has on innovative soil treatment technologies. There is also a bimonthly newsletter called "Tech Trends" that features three projects on soil or groundwater remediation. A looseleaf notebook is also being prepared, including major reports on new technologies organized by topics such as soil washing, vitrification, and bioremediation. Bioremediation has become so important that special bulletins are now being issued. Over 130 field projects, including 30 that are not Superfund-related, have been identified, and the phone numbers of project officers are included so that interested parties can discuss work in progress. The EPA also sponsors international conferences that focus on major environmental topics and collaborates with other agencies and environmental groups on nationwide satellite seminars on topics such as bioremediation. There is also a computer network-based dial-up bulletin board system that allows interactive discussion among those working in particular environmental areas. Another project is a survey to be issued on computer disc of U.S. vendors of innovative technologies.

Because of the huge remediation problems still to be addressed, the EPA recognizes the need for wide distribution of the latest information. One example is the 1.8 million underground storage tanks that have been catalogued—about 90 percent containing petroleum—with an estimated 10–20 percent of them already leaking. There may be only up to 5,000 cubic yards of contaminated soil at each site, making for a relatively small project, but there are so many such sites that a large number of construction firms will need to know effective methods for proper treatment. The Department of Defense (DOD) has more than 7,400 sites that need treatment, and DOE has a number of hazardous waste sites of 10–100 square miles in size, making them megaprojects.

Dealing with such work will require even the largest construction companies to acquire skills in multiple areas. Even without entering the environmental marketplace, however, construction companies will have to add specialized know-how. In any future construction, environmental considerations will be important throughout the project. Applicable regulations must be met and permits obtained, often with public oversight. This adds new layers of complexity to any construction work, with more opportunities for delay and potential disputes with public authorities. Experience in dealing with environmental agencies, along with skill in streamlining the process to minimize delays, becomes an important asset. Effective computer-based data management is also critical for handling big projects smoothly.

Achieving source-control objectives requires much more than meeting environmental regulations and obtaining permits. There must be close involvement with the client's operating specialists to come up with appropriate methods for eliminating waste, treating it for reuse, or converting it into useful by-products. Consulting/engineering and construction firms can sometimes offer packaged environmental solutions, but more often modifications will be needed to adapt to the needs of a particular client or an individual facility.

There is another good reason why environmental firms and construction companies that operate in this market need to expand their capabilities in the source-control area. Currently the trend is being driven by a few clients looking to improve their environmental image, cut costs by eliminating waste, and perhaps find new sources of profit by properly treating process by-products. But, in the future, that could change as many more clients look toward source-control solutions because of regulatory shifts.

So far regulations in the United States have served to push funds into waste management, pollution control, and remediation. It is likely that within the next five years the next level of regulations, in the United States and in other nations, will put more emphasis on cleaner product and process technologies. Regulations aimed at boosting source control will become more forward looking and technology transforming rather than just continuing to refine existing waste-management practices. That means that companies that have already made investments in various environmental sectors, such as pollution control equipment and other end-of-pipe control technology, may begin to see some markets decline. Construction companies with skills applicable to the source-control approach will be likely to find plenty of new growth opportunities involving such work as setting up new process and manufacturing lines, particularly by developing turnkey systems aimed at source reduction and offering skills and knowledge applicable to reducing or eliminating waste and pollution within various industries.

Even if a construction company determines that it does not wish to enter the environmental market and chooses to subcontract any pollution-abatement or source-control work on its projects, it will still have to meet stiffening environmental requirements for its own work. Pressures will increase for the use of environmentally acceptable materials, reduction of construction waste, and recycling of building materials. Rules regarding construction processes and materials have already become a major issue in Europe, and the trend is bound to move into the American market. Again, those construction firms that take a lead in introducing greener practices may find this approach will give them a competitive advantage with some clients.

A growing number of construction firms are entering the environmental market through specialized divisions or subsidiaries, but the trend toward source control is bringing environmental considerations right into the planning phase for any new or upgraded facilities. This means that construction contractors will have to remain up to date on techniques for reducing or eliminating waste and pollution so that they can work with their clients to achieve these goals. Skills gained may be applicable to expanding global environmental markets, but construction firms will have to adapt to international business methods geared to information systems and telecommunications advances. Partnerships and other types of alliances with overseas firms may also be needed to break into these markets.

Source control is just the start of a broader trend toward sustainable development. It is clear that sustainable development cannot be achieved using the methods of the past. New methods involving new technologies will be essential, but under the present regulatory climate numerous barriers hinder the development and application of new technologies. Efforts are already under way to lower some of these barriers, so that construction companies alert to opportunities in environmental markets and in source-control methods can be in a position to prosper in the future. Some of these efforts will be described in more detail in Chapter 4.

CHAPTER 2

INDUSTRIAL PERSPECTIVES AND INITIATIVES

INDUSTRIAL PERSPECTIVES AND INITIATIVES

Industrial clients of construction contractors represent a major portion of the total business. Construction work for the industrial segment of the market encompasses the building of factories, processing plants, storage and materials handling facilities, and office buildings (including corporate headquarters).

Environmental concerns have played a steadily expanding role in construction for the nation's industrial sector as regulations have increased and broadened in scope. Traditionally, industrial firms have considered environmental protection as an externality that, at best, should be dealt with by the public sector. But in recent times the influence of the environmental movement has become much more invasive, requiring much more attention and investment by industry to reduce and treat pollutants and other waste. While industry's early approach was reluctant compliance to environmental regulations, now there is a major transformation in attitudes taking place. Industry's focus is shifting from merely meeting regulations toward finding ways to better control, reduce, or even eliminate pollution. Industrial firms are looking for more help and ideas from contractors as they seek ways to modify their operations to satisfy the emerging requirements for sustainable development. These companies expect engineering and construction contractors to be knowledgeable about environmental trends and are looking to utilize this expertise to help them find effective solutions.

This chapter provides the historical background for current environmental trends and reviews how attitudes have been changing, particularly within industry, as the "green" movement gains momentum. It includes discussion of specific events and major studies that have influenced thinking about the limits of the world's resources and the ability of the earth to absorb pollution and waste.

In the early part of the twentieth century, during a period of increasing urbanization and industrialization, environmental neglect was common. Public concerns about deteriorating environmental quality eventually led to regulatory legislation and, as the rules tightened, industry was forced to change its approach from neglect to begrudging compliance. Executives generally considered environmental controls as a growing cost being forced on their companies by factors external to the primary business. In the absence of regulations, even if a firm invested in pollution-control or waste-management equipment, it might become uncompetitive in its industry because of the added costs.

Attempts are now being made to integrate environmental considerations into conventional economic systems, although with some difficulty as will be discussed later. This chapter will review these developments, along with some examples of successful efforts by a few companies to actually profit from their environmental efforts. The changing outlook on the environment that has resulted will shape the regulatory climate that builders will face in the future as well as the customer expectations that contractors will need to satisfy.

FROM NEGLECT TO PROTECTION

Until the middle of the twentieth century, the environment's capacity for absorbing all kinds of waste was generally considered to be boundless. In this view, natural processes would gradually dissipate any pollutants, whether disposed of in the water or the atmosphere or buried in the earth. Nature, given time, would also heal any wounds left behind by operations such as mining and lumbering once the target resources had been extracted or harvested. Natural dispersion of noxious gases from industrial processes by the atmosphere could be given a boost by building smokestacks higher. Rivers, streams, and the oceans seemed ideal for washing away everything from used solvents and oils to sewage sludge, and sometimes even radioactive debris. Remaining waste could be disposed of in sanitary landfills, typically located at the edge of towns or cities. To prevent groundwater contamination in these dumps, any toxic waste could be poured into 55-gallon drums before burial.

Waste disposal was considered to be a public responsibility to be paid for by taxpayers and property owners. Industrial firms often hired outside contractors to dispose of their waste, and they left it up to these outsiders to comply with applicable regulations. Neither producers nor consumers were considered to be responsible for environmental degradation (except perhaps for flagrant abuses, such as releasing toxic substances that could contaminate public drinking supplies), since any damage was assumed to be only temporary anyway. Instead, society suffered a loss of amenities, such as woodlands, or a decline in the quality of life in the form of foul air and polluted rivers and harbors, without having these losses factored into the cost of production or the prices for goods or materials.

This period, which extended from the beginning of the industrial revolution through the 1960s and into the 1970s, might be considered an era of environmental neglect. Since industry had little financial accountability for the effects of pollution and waste disposal, it was convenient for industrial firms to embrace the concept of an environment with limitless capacity to absorb waste products. The costs of cleanup or other residual damage resulting from poor environmental practices were the responsibilities of others external to the firm, so these matters were given little consideration in planning for facilities and operations. Any resources devoted to waste management and pollution control were considered to be outside the main business of the firm, so these expenditures were viewed as similar to other secondary functions such as maintaining buildings and grounds or shipping goods. Top management in most firms ignored environmental matters, and, even in those companies concerned about being environmentally responsible, the function was typically relegated to secondary departments such as plant maintenance or janitorial services. Governments themselves were some of the biggest polluters, with localities dumping untreated sewage into waterways while leakage from town dumps seeped into water supplies. The federal government was a major polluter as well, with the military in particular dumping or burying toxic materials and, in some cases, nuclear wastes on or near bases, not just in the United States but around the world.

The limited scale of population density and development up through the early 1900s also helped to foster the view of the environment as an infinite sink. The transition from largely agrarian to industrial societies took place over many decades in the more advanced countries such as the United States. Metropolitan areas expanded steadily, providing essential infrastructure for business and industry, including facilities such as airports and highways as well as concentrations of suppliers of parts and services. Since this urban growth was gradual, there was time to cope with at least some of the environmental impact of the resulting concentration. Municipalities in the United States began to prohibit the belching black smoke that had begun to engulf some highly industrialized areas such as Pittsburgh, Detroit, Cleveland, and Gary, Indiana. The capital needed for waste treatment or disposal came from the economic success of the industries creating pollution or waste, and any associated costs could be passed on to customers. Lumbering and mining operations, however, remained relatively unimpeded by environmental concerns because they were carried out in remote or sparsely populated areas, and only small numbers of people were directly affected by practices such as the clear cutting of forests or open-pit and strip mining. In the western United States, piles of tailings from the mining of metals such as copper, silver, lead, and molybdenum were eventually buried or poured into tailing ponds.

Still, the stirrings of environmentalism had begun by the start of the twentieth century, with the conservation ethic preached by prominent lovers of the outdoors, such as Theodore Roosevelt and John Muir. As president, Roosevelt became a strong proponent of national parks, believing that they had to be protected so as to preserve their natural beauty. He was influenced partly by Muir, a brilliant photographer whose pictures captured the majesty of mountains, deserts, redwood forests, and the essence of places such as Yosemite National Park in California.

By midcentury, environmental requirements had begun to take shape for the by-products of some processing operations, such as for solvents used to extract metals from ore or the residues of pulping and papermaking. Many growing communities also began to address environmental problems that had long been ignored. By the 1950s, most open dumps were being converted to sanitary landfills in which garbage and debris from both consumers and industry were covered by soil, generally within 24 hours. The open burning common in earlier times was prohibited in these improved landfills.

ENVIRONMENTAL PRESSURES BUILD

Through the 1960s and 1970s, increasing public concern about the environment in the United States began to build pressures for much stronger regulation. In 1962, Rachel L. Carson's book, *Silent Spring*, attributed much of the decline of bird populations and some other types of wildlife to the widespread use of pesticides and agricultural chemicals, particularly DDT. This book is often cited as the catalyst for the rising public concern about environmental issues in the United States that has now spread all over the globe. Along with the antiwar movement of the 1960s came a growing disaffection with "the establishment." Polluting practices of industry became one convenient target for increasingly militant opposition to "the system." Ralph Nader, the consumer rights crusader, spoke at universities and forums all over the country, calling attention to the widespread practice of flushing industrial chemicals into streams and rivers and provocatively suggesting that "industry needs some toilet training." Indeed, fish and other aquatic life *did* almost disappear from the waters of Lake Erie due to the cumulative effects of chemicals and other pollutants pouring into it from industries and communities along the streams and rivers that fed the lake. Jacques Cousteau, the French oceanographer, reported that even the oceans were beginning to suffer from the cumulative load of waste fuel oil and other pollutants being dumped into them. Trees began to die on mountain tops in the northeastern United States and eastern Canada, and smaller lakes began to suffer a loss of fish life. Environmentalists charged that the damage was due to acid rain created by sulfur and nitrogen compounds emitted primarily by electric utilities and factories in the U.S. industrial heartland. Concern even began to rise about the erosion of public statues and architectural landmarks, also attributed to acid rain. Trace metals began to show up in western watersheds, carried into groundwater as the winter melt seeped through billions of tons of mine tailings.

Even though concerns about the environment were rising in the intellectual community and with some of those directly affected, these issues did not have much impact on the national political agenda through the 1960s. They were rarely mentioned in the 1968 presidential campaign, for example, when there was heavy emphasis instead on the Vietnam War and the problems of urban poverty. In 1969, however, an oil-rig blowout in the Santa Barbara Channel leaked thousands of gallons of crude oil just off some of Southern California's most pristine beaches. This widely publicized incident illustrated to Americans the fragility of the relationship between man's activities and the environment, and showed how ill-equipped the nation was to deal with such disasters. Shortly after that incident, Wisconsin Senator Gaylord Nelson proposed the idea of "a nationwide teach-in on the environment" during a lecture at the University of California at Berkeley. This helped start a trend toward an organized environmental movement, which led to Earth Day in 1970 and the creation of the Environmental Protection Agency in 1972, which can be viewed as marking the transition between the conservation ethic of the first half of the twentieth century and the activist era of environmental protection based on command-and-control regulation that followed.

Disputes between industrial firms and nearby residents were infrequent, and when friction did occur there was little in the way of environmental legislation to protect citizens. It was difficult to assign legally direct responsibility for environmental damage to offending industries, and enforcement tended to be casual even for the few restrictions that did exist. The main concern of most companies was to minimize the costs of disposing of waste and effluent while hiring lawyers and using the courts to fend off occasional environmental challenges.

By the 1970s, due to growing U.S. affluence and industrialization, many urban regions had expanded to engulf factories that had once been on the outskirts of town or even out in the country. Higher-density living also generated far more garbage and sewage, along with exhaust gases from growing numbers of automobiles and trucks. All of these factors added to the load on the environment. As population densities increased, many more citizens were effected by mounting environmental problems, including noxious fumes from urban smog and pollution that fouled rivers, harbors, shellfish beds, and especially beach areas. The terms "ecology" and "ecosystems" increasingly entered the public debate over the environment, reflecting a shift in attitudes. These terms referred to the complex and often poorly understood relationships between all living creatures and the role of the natural world in supporting life. Scientific studies of ecosystems had traditionally focused on the natural world, such as the rain forests or oceans, but in the environmental debate ecological concepts were broadened to include mankind's relationship to other species and to nature, including the impact of human activities on the planet and its ability to support life. Perceptions began to rise that pollution and waste were being generated in greater quantities and at faster rates than could be effectively dissipated. The traditional view that the environment could readily absorb all the waste resulting from human activities began to dissolve, adding urgency to the rising clamor for greater curbs on industry.

INDUSTRY VERSUS ENVIRONMENTALISM

Public pressures for increased environmental regulation were generally resisted by industrial leaders. But gradually, as regulations forced changes in practices that had led to some of the environmental abuses of the past, there was a transition to a period of begrudging compliance on the part of industry. Requirements for pollution control equipment and increasingly stringent restrictions on waste disposal were viewed by most executives as costly and sometimes excessive measures, often shaped more by politics than by unassailable scientific evidence. Many firms accustomed to operating without environmental constraints bridled not just at the added expense, but also at the need to change their methods and processes. When pressed to meet new environmental rules, these firms often fought regulators' demands in court, putting their resources into legal defenses and paying fines rather than investing in new equipment and changing their firms' ways of doing things. Other companies, witnessing the troubles of those who failed to meet environmental standards, set up special departments and charged them with doing whatever was necessary to meet applicable regulations, but with the minimum possible investment. Lobbyists hired by industry joined trade associations in either opposing new environmental legislation or urging moderation in any regulatory provisions.

Most companies made little effort to reduce or eliminate waste or pollution at the source. This attitude was consistent with the basic approach of regulators, whose efforts were devoted to controlling end-of-pipe discharges and dictating control technologies rather than encouraging elimination of waste and pollution altogether. Under such a regime, management still considered waste disposal to be external to the main business of the firm, generating overhead expenses that had to be added to the costs of doing business. Separate departments set up to deal with environmental compliance typically operated in isolation from the profit-making centers of the corporation, and meeting regulatory requirements was considered to be a necessary evil that had little to do with day-to-day operations.

Few companies chose to become leaders in the environmental movement, innovating and going beyond what was legally required. One good reason was the perception that the higher costs incurred in trying to outdo other firms in controlling waste and pollution would make a firm noncompetitive within its industry. In fact, one of the major justifications for the command-and-control approach to regulation was that all competitors had to be forced to make investments to meet the same environmental standards simultaneously so as to equalize overhead costs. It was only after considerable experience with this process that it became apparent that the actual costs for meeting similar standards, even within the same industry and within the same geographic region, could prove in practice to be far from equal. Still, some companies, seeing the value of being viewed by consumers as environmentally sensitive, did make at least token gestures to move beyond regulatory requirements, while at the same time seeking widespread publicity for their environmental efforts.

RISING CONCERN ABOUT ECOLOGY AND INDUSTRY'S ROLE

Through the 1970s and 1980s, public concern about the environment intensified dramatically. Incidents such as the forced evacuations of an entire town at Times Beach, Missouri, and a neighborhood near Love Canal in Niagara, New York, sparked national outrage. To suppress dust, streets in the Missouri town had been sprayed with waste oil from a local chemical plant, which proved to contain dioxin, a highly toxic carcinogen. At Love Canal, homes and even a school had been located on top of a former hazardous waste dump, creating serious health hazards when the barrels of toxic chemicals buried there began to leak, contaminating drinking water and even bubbling to the surface in some spots. Hooker Chemical Corp. had buried its toxic waste in the dump for some twenty-five years, following practices that the company claimed had been in accordance with the applicable regulations of the time.

Widely publicized cases like these helped to focus national attention on the folly of past methods of waste disposal and pollution control. Even well-managed sanitary landfills lacked subsurface barriers to prevent toxic substances—perhaps from discarded materials such as household cleaning agents, used automobile batteries, waste oils, and solvents—from leaching into water supplies. Moisture seeping through the ground also caused metal drums and barrels to rust and leak, eventually releasing their toxic contents. Waste matter decaying underground in landfills also generated methane gas which created another hazard. On occasion, fires, explosions, and even some deaths were attributed to this colorless, odorless gas (commercial natural gas has an odor-causing substance added to it so that leakage can be detected).

Legislation was passed throughout the 1970s and 1980s to address these problems, and the EPA was charged with developing programs for the remediation of the legacy of the past. The Superfund program began in 1980 to clean up potentially harmful toxic materials that had accumulated in thousands of dump sites around the nation. Tougher curbs, particularly for hazardous materials, were instituted to prevent further contamination of soil, water, and the atmosphere. At the time this program was being established, neither government officials nor corporate executives had any idea of the magnitude of the abandoned waste site problem. In 1979, an EPA study concluded that there might be between 1,200 and 2,000 hazardous waste sites that might pose threats to human health and the environment. They estimated cleanup costs at anywhere from $3.6 billion to $44 billion (the wide range indicated the lack of solid data on the actual dimensions of the problem). Many legislators begrudgingly voted for the program, expecting it to be of relatively short duration. Later the EPA's Hazardous Ranking System list grew to more than 26,000 sites, not including the 1,200-plus sites on the National Priority List. The General Accounting Office (GAO) has predicted that the list could grow to 368,000 sites when a more comprehensive inventory is taken. And the Office of Technology Assessment estimates that just cleaning up the priority sites could cost more than $500 billion during the next 50 years!

Other environmental rules were strengthened as well, although many did not include the sort of enforcement "teeth" that would have made corporate executives sit up and take notice, as did the Superfund legislation. For example, some required the landscape to be restored after strip-mining operations. Others attempted to force communities to cut back on air pollution, and still others aimed to clean up water supplies. But even these more stringent measures, which have been evolving during three decades of experience, are now being recognized as only slowing the process of environmental degradation. This discovery comes as no surprise to those who have long followed the still-continuing debate over environmental and conservation policy making.

Many industry and trade representatives argued that often the costs of these programs far outweighed the benefits when determined by traditional economic assessment measures. Opponents of environmental action were convincing in their arguments, partially due to the shortcomings of traditional accounting methods when trying to deal with the diverse, intangible benefits that characterize much of the environment. This discrepancy forced many to reexamine the underlying economic principles that formed the basis for traditional analysis.

THE TRADITIONAL ECONOMIC MODEL

The traditional economic model on which the economies of the United States and many other industrial nations are based fails to deal with environmental issues. While open markets are supposed to induce steady economic progress toward greater productivity and growth, there is a lack of natural incentives encouraging innovation and technological solutions for pollution control and waste management. Without some sort of government intervention in the market, the ultimate environmental impact of actions taken by producers and consumers is not factored into the cost of goods. In fact, in many cases the marketplace even offers perverse *rewards* to those creating the most damage to the environment. Excessive packaging may win far more customers than that made from minimal amounts of recycled materials, for example. Manufacturers often locate plants in regions not as sensitive to pollution, such as Mexico, where environmental rules are less severe and authorities may be willing to excuse environmental degradation to attract job-creating factories. This practice can either drive more environmentally responsible competitors out of business or force them to move operations to lenient areas as well. The greatest economic rewards tend to go to those who are most effective at externalizing the environmental consequences of their actions.

There are a number of straightforward reasons cited for the shortcomings of the conventional markets in dealing with the environment without the intervention of government. The basic economic model postulates that in open markets, with individuals free to negotiate prices while buying and selling products and services, there is no need for centralized planning and control because marketplace competition will ensure an efficient allocation of resources. The model assumes that "rational actors" will make coherent and logically consistent decisions, choosing options according to the preferences and opportunities allowed by society. The available opportunities will be limited by resource availability, technology, and public morality (which dictates acceptable and unacceptable behavior). Opportunity costs, or prices, will then determine choices by individuals, and, conversely, individual choices will influence prices. In theory, for such a system to work properly, each economic actor needs to be fully informed of all prices and of the consequences of every transaction, and no one can manipulate the prices others would charge. Also, rights to all property must be absolute and completely transferable among parties without transaction costs.

If markets operate within the criteria specified by this model, with producers and consumers operating in their own self-interest, the theory concludes that their collective actions should bring outcomes that are beneficial to all. Prices provide the signals to producers and consumers that guide them in adjusting production, distribution, and consumption for a wide variety of goods and services so as to maximize the gain of each participant. The outcomes resulting from the cumulative transactions of these rational actors, all of them looking out for their own selfish interests within "perfect" markets, will be for the greater good of society.

In practice, however, conditions in real markets often fail to meet the assumptions and constraints on which the theory of market exchange is predicated. The determinants of private costs, for example, may vary widely from those for public costs, and the market cannot easily reconcile the two. A firm may incur little cost by dumping waste gases into the atmosphere. But if these pollutants contribute to the enhancement of the greenhouse effect, the eventual cost to the society as a whole resulting from the accumulation of such gases may be high indeed. In an efficient market, meeting all the theoretical criteria for property rights, the marketplace will automatically allocate a balance of rewards and penalties. If costs exceed benefits, no action will be taken. But if all costs are not realized, with some of them remaining external to the calculations of individual decision makers and the marketplace, then the market will not serve society efficiently. It is these "externalities" that are often cited as the most important cause of environmental difficulties. The result is that some agent or agents in the market will benefit at the expense of others.

Aside from the discharge of effluents by industry, there are many other examples of external costs, including pollution from industrial and commercial activities; agricultural runoff; changes in surrounding property values due to the siting of an unwanted utility, factory, or landfill; or escalating rents when a large commercial interest moves into an area and drives up prices. Sometimes the effects of externalities can be positive rather than negative, such as increasing property values due to development of a nearby park or the cleaning up of a formerly polluted river. Only when the social costs of actions become internalized so that they are naturally taken into account in making economic decisions will the interests of all the individual actors and the overall society coincide.

Traditional economic models also have trouble dealing with common property that is not controlled by any single party. This includes our natural surroundings, such as the air, rivers and oceans, wildlife, and scenic public parks, as well as shared resources, such as ocean fisheries or even massive oil or gas reservoirs in some cases. If there is unrestricted access to common property and resource extraction costs are low, the resource becomes overexploited because no individual has an incentive to conserve. If an individual had complete control over a declining resource, he or she would practice conservation because benefits would be derived from the long-term income derivable from a scarce commodity. But if individuals must compete for the used of common property, their incentive to save is lost, and the individual parties will try to maximize their own return before the resource is depleted or ruined.

A similar distortion of incentives under traditional capitalism occurs when someone owns a property with high social value but low market value, such as a wetlands. In this case, the benefits of the property do not accrue to the owner but to others. The wetlands area may help control flooding for those downstream, provide spawning grounds for fish that then migrate elsewhere to be caught by fishermen, and filter and biodegrade contaminants in river water to the benefit of downstream properties. As a result, the owner has an incentive to maximize his or her own benefit by draining the wetlands and converting it to farmland or building lots to the detriment of society.

Another defect of the traditional theory of market exchange in dealing with the environment is that there is no market for valuable public goods such as the clean air, rivers, and oceans, and a beautiful landscape. They are not "traded." Without a marketplace, it is difficult to assign them prices, so that they might be efficiently allocated either by the market or by external influences, such as government actions. Debates rage in the safety and health arena, for example, over how to assign a value to a human life. Unfortunately, contention over such matters, including many environmental issues, tends to be reduced to a group of eloquent and intelligent parties with opposing interests trying to argue each other out of their moral and philosophical convictions (similar defects also apply to Marxist doctrine). There are some aspects of the environment, such as a scenic view, that cannot easily be fitted into an economic context through some sort of pricing mechanism. Some sort of consensus may have to be reached among a large public in order to reach rational methods for setting rewards and penalties to try to maintain a desired level of preservation.

Nevertheless, there have been a few cases of relatively successful efforts to assign monetary values to some environmental amenities, such as clean air and fishable waters. This approach has helped protect some scenic rivers in California and wetlands along the Mississippi River, for example. Recently the state of Oregon has begun to consider assigning dollar values to such amenities as forested mountains and pristine beaches. Such valuations should go well beyond pitting the estimated $30–40 a day spent by visiting tourists against the income of timber interests, according to Edward Whitelaw, a University of Oregon economist. Dollar figures should also include the quality of life afforded to urban workers and others who can enjoy close-by wilderness areas that have been gradually denuded by logging operations. In California, policy makers recognize that many factors must be considered in determining benefit/cost payoffs for tougher environmental rules. Health care costs have been rising rapidly, for example, and respiratory diseases are more common in areas with higher smog levels. Higher costs for less polluting vehicles and factories may be returned to the society later by a reduction in medical costs and in time lost from work due to health problems.

Difficulties also arise when a single party (a monopoly) or perhaps a small group of companies (an oligopoly or trust) has the power to influence the price or supply of assets in the marketplace. With many buyers and sellers, natural competition tends to produce efficient allocation when there is restricted access and relatively high extraction or development costs. By severely limiting competition, pricing likely will be inefficient and not in the best interests of society. This problem is sometimes cited, for example, in the case of tough restrictions for new landfills, including requirements for impermeable liners to prevent leaching into water supplies, systems for removing methane, and long-term liabilities for operators. The result may be that, by early in the next century when most present dumps have been shut down, there will be only a few very large landfill sites controlled by perhaps one or two large corporations with the ability to manipulate dumping fees. Command-and-control approaches dictating "best available technology," or requiring approval of plans by regulators, can similarly result in narrowing suppliers to a very few or even a single vendor. Constricted markets can also result from patents or even from the success of a vendor in cornering a marketplace. One factor contributing to limited competition in environmental markets is the difficulty in assessing risks so that potential costs can be estimated. Vendors must often assume long-term liability for potential damages in areas where there is great uncertainty, including the potential for a costly disaster in the case of applying new technology to dispose of highly toxic wastes, for example.

An economist might argue that, if a monopoly is created that charges excessive fees or prices, this would produce strong incentives either for competitive suppliers to find a way to enter the market or for the development of alternate technologies. In the case of the environment, in some cases the alternatives, rather than being "improvements," have included midnight dumping and other surreptitious methods of waste disposal or regulation avoidance. The search for new technologies or better legitimate methods for controlling or even eliminating pollution and waste could stretch over quite a lengthy period. During this interval of economic adjustment, while innovators with better technology or more efficient methods try to challenge the market control exerted by a monopoly and officials seek more effective ways to enforce regulations, severe and perhaps irreversible harm could be done to the environment by those seeking ways to avoid paying monopolistic prices.

FAILURES FRUSTRATE POLICY MAKERS

The failure of normal economic mechanisms to adequately protect the environment remains of great concern to policy makers. Most environmental legislation in recent times has attempted to sidestep marketplace economics by mandating the addition of expensive, end-of-pipe discharge technologies to deal with waste after it has been generated. This approach virtually assured confrontational interchanges between regulators and industry, and the outcomes have often been less than optimum. For example, the separate rules for gaseous emissions, liquid discharges, and solid waste have been steadily toughened, in a segmented fashion, by adding new amendments to existing laws. In response to this piecemeal approach, companies have often opted to alter their waste by-products to whatever form has the least stringent requirements at the time. Smokestack scrubbers, for example, convert the contaminants in gaseous emissions into liquid wastes. These liquids, in turn, can be filtered to create a solid sludge containing the original contaminants. Such convertibility affords the polluter the option of choosing the least troublesome, minimum-cost approach to disposal without even considering eliminating wastes at the source.

Actions such as these frustrate policy makers. Instead of new regulations achieving a hoped-for reduction or elimination of pollutants and waste, they lead instead to a sort of dog-chasing-its-own-tail response by industry. When challenged by such criticism, industry can counter by charging that regulators often fail to do a careful benefit/cost analysis before toughening regulations. Achieving an additional 5-percent reduction in some pollutant, for example, can cost far more than it did to reach a 90-percent reduction in the first place, and there can be a huge cost differential between restoring the landscape to a livable state and returning it to its original pristine condition. As rules tighten, mounting costs for compliance can reduce the competitiveness of American firms in global markets while encouraging some companies to move production offshore to areas with less stringent environmental rules.

Even as the costs of achieving tougher environmental rules have escalated, accompanied by confrontation and growing animosity between regulators and industry, progress toward a cleaner environment has been slow. Policy makers would like to see environmental planning become a natural part of corporate decision making, rather than a begrudging afterthought. They want industry to give this area consideration equivalent perhaps to that applied to choosing production equipment or selecting a plant site. One approach is to force environmental costs into corporate financial decision making in such a way as to provide rewards to those firms with the most effective programs and penalties to those that lag. To accomplish this "internalization," there are now efforts to steer policies toward creating incentives for meeting environmental goals. Some of these are direct incentives, such as:

- charging fees or taxes for each unit of pollution discharged
- achieving an overall level of pollution control by allowing firms to buy and sell pollution permits
- using deposit-refund systems where surcharges are added to potentially polluting products so they will be returned for reuse or recycling

In the case of toxic substances, the Project 88 Report, prepared as a policy guide for President Bush in 1988, suggests that charging an emissions tax or discharge-end penalties for waste products encourages "midnight dumping." Since every shipment from a plant cannot be monitored, small amounts of material might also be disposed of by hiding the hazardous material within loads of other material. Present rules requiring plants to obtain permits for discarding hazardous waste also encourage such surreptitious activity. A study of industry practices revealed that roughly half of such substances as degreasing solvents and lubricating oils could be filtered and reused, and yet this was done only rarely. Rather than a penalty at the discharge end, the Project 88 Report recommends encouraging conservation and reuse of hazardous materials by adding a front-end tax or a deposit that would be refunded when the containerized material is turned in for recycling. The user would be held responsible for assuring that the proper substance was turned in to the collection station. This would encourage users to minimize losses through evaporation or emissions, thus also helping to reduce the release of harmful precursor gases. Such fees would also encourage a search for less hazardous substitutes.

There are also indirect market incentives that do not assign a direct cost to waste but instead serve to limit the available options for dealing with them. An example of this approach is recent legislation establishing a retirement agenda for existing landfills while setting tough requirements for creating new ones. This process is bringing steep increases in disposal costs, thus forcing industry to look at alternatives such as source reduction or processing of waste so it can be reused.

Since it is always difficult and sometimes impossible for regulators to assign correct evaluations to incentives or penalties, policy makers are recognizing that, the best approaches will allow industry some leeway in choosing the means to desired ends, while providing rewards to those that succeed in finding ways to achieve better environmental results at less cost. According to economic theory, a new approach to environmental protection based on such principles would induce more innovation while speeding up progress toward environmental goals through the hard-headed competition of the marketplace.

Unfortunately, studies of actual cases in industry where cost savings could be achieved through better environmental practices show that, even when savings could be substantial, there is frequently little effort to change existing practices. Energy conservation by industry, for example, offers great potential for cost savings while also reducing the generation of greenhouse gases. But with energy costs buried in corporate overhead and considered by management as extraneous to the main order of business, these matters receive little top-level attention. Similar lost opportunities were found in areas such as recycling scrap materials or filtering and reusing solvents.

Some policy makers hope eventually to induce industry to go beyond the narrow dollars-and-cents approach of minimizing compliance costs. They would like to see environmental planning move into the corporate agenda as an element of good business and civic responsibility. Achieving this appears to be a formidable task. Values such as living in harmony with nature are more reflective of some Eastern philosophies than the hard-headed, bottom-line orientation of most American business executives. But it is difficult to find ways to encourage progress toward environmental objectives based simply on the dollars-and-cents approach. More ethereal measures may be required, such as the greater good of society, esthetic sensibilities, and even morals and ethics. Mechanisms are being sought, however, that do, in effect, translate some of these values into economic decision making by businesses and individuals.

Developing policies that achieve a reasonable balance between nuts-and-bolts economics and these more elusive objectives requires sorting through myriad alternatives while using evaluations that, by their nature, require highly subjective, sometimes even philosophical judgments. In fact, environmental concerns do appear to be starting to penetrate some corporate boardrooms and executive suites.

U.S. CORPORATIONS FORCED TO REVIEW ENVIRONMENTAL POLICIES

In the United States, the growing array of environmental legislation has forced corporations to look more closely at their policies and operations. Concern was particularly heightened by legislation assigning long-term liabilities to the generators of pollution and waste. Even when disposal was done legally years ago, corporations have been held responsible for environmental damage that occurred later. While paying fines and court costs was sometimes cheaper than meeting environmental rules in earlier times, recently the penalties handed out by the EPA and by the courts have gotten much tougher. While criminal fines for environmental crimes were less than $2 million a year in the United States in 1985, by 1989 they had increased to more than $13 million. The amounts have continued to escalate. In 1991 Exxon Corp. alone paid a $100 million criminal fine as a result of the *Exxon Valdez* oil spill in Alaska. Companies not only must pay fines and legal costs but also must cover cleanup costs under the Superfund program. In a few cases, executives have been held personally responsible and have gone to prison as a result of negligence. Overall, the Justice Department has handed out an approximate total of 348 years of jail terms for environmental crimes since 1983. In the 1991 fiscal year, the EPA referred a record 81 cases to the Justice Department, resulting in 125 indictments and 96 guilty pleas. These cases yielded $14.1 million in fines and 550 months of prison time. Most of these cases involved violations of RCRA regulations, which will be described in the next chapter.

As corporations review their environmental policies, some executives are learning for the first time the huge costs of the waste products of their operations. This process is leading to the internalization of at least some of their firms' overall environmental costs as active programs get under way to cut back or eliminate waste treatment and disposal. Reducing waste, which can be defined as anything that does not contribute to meeting customers' needs, is by definition a major tenet of the total quality approach. Not surprisingly, this process has resulted in fairly profitable projects in many cases. If managers view the process of environmental management as simply another source of competitive advantage gained by the reduction of waste, they often can see processes in new ways that can lead to important advances. This approach, best described as the Total Quality Management (TQM) philosophy, has recently become an avid pursuit of many industry leaders engendered by the national

Malcolm Baldrige National Quality Award program that began in 1988. Environmental stewardship is, in fact, part of the criteria used to select Baldrige Award winners.

As multifunctional teams reexamine industrial processes, many firms have eliminated, or at least minimized, pollution and waste and actually reduced costs while cutting the risks of future liabilities. Products already in production can be redesigned to make reuse or recycling feasible. As the team approach becomes integrated into American corporations, environmental considerations are being factored into decision making from the outset of a project, whether for new products or whole factories. The continuous improvement methods underlying total quality programs work by setting very tough goals and then finding ways to achieve them even though they initially seemed unreachable. As reuse and recycling of former waste products expands, companies are finding that looking at a total process often introduces new options that had never even been considered before. Improvements can be dramatic, but then new, even tougher, goals are set, and the teams search for ways to make still further progress.

Not surprisingly, some of industry's leaders in improving their environmental performance are also among the nation's largest polluters. Firms such as these have the most to gain by learning to better manage their waste and pollution. E. I. DuPont de Nemours, for example, recently announced a $500 million capital improvement plan at three chemical plants in the Carolinas that will cut atmospheric emissions by 60 percent while boosting production by 20 percent. Large companies such as Dow Chemical, Amoco, Monsanto, and DuPont have voluntarily agreed to reduce pollution from 17 toxic chemicals by 33 percent in 1992 and by 50 percent by 1995. They are among 600 firms approached by the EPA in 1991 for voluntary participation in an Industrial Toxics Program better known as the 33–50 program because of the targets set for reduction. The Chemical Manufacturers' Association has asked its 185 member companies to commit to a set of guiding principles for management, including dedication to pollution prevention. Minnesota Mining & Manufacturing (3M), Polaroid, BASF, Exxon, General Electric, Goodyear, Occidental Petroleum, Reichold Chemicals, Texaco, Texas PetroChemicals, and Procter & Gamble are among a number of other large companies, either in the chemical business or large users of chemical products, that are actively pursuing strategies to drastically cut waste and pollution. Ford Motor Co. is pressing all its vendors to come up with recycling plans for materials they supply for Ford's automobiles and trucks.

Nevertheless, much of industry still views compliance with regulations as a nuisance and a cost center, and many executives remain convinced that environmental protection stifles economic growth. A few corporate leaders who are positioning their firms as pioneers in "the greening of industry" have been trying to sway the views of fellow executives because they see rising public indignation as a threat to their future operations. At the same time, they are providing some concrete examples of how "industrial ecology," as it is sometimes called, can actually pay off.

A recent speech by A. F. Jacobson, chairman and chief executive officer of 3M, summed up the major reasons why he believes industry's attitudes should change. Since the trend toward greater concern about the environment will continue for the foreseeable future, he suggested that industry might find it advantageous to set tough goals for itself that extend even beyond regulatory limits. In brief, Jacobson's reasoning is as follows:

- If industry waits for regulations to be imposed, there's an excellent chance they will be more costly, less flexible, and less efficient than what industry can do on its own to reach even tougher, self-imposed goals.
- If goals are set correctly, industry should also reduce overall production costs because pollution is waste, which is bad not just for the environment but also for productivity.
- Industry can reduce the risks of future liabilities, including some that may not even be known yet.
- Reformulating products and revising processes to make them more environmentally friendly can often result in products that are more marketable as well.
- Better environmental performance enhances a corporation's reputation and helps build and maintain a reservoir of good will with customers and the public. This is especially important within the communities where the company's operations are located.

THE 3M POLLUTION PREVENTION PROGRAM

In 1975, 3M launched a program it calls 3P, or Pollution Prevention Pays, which is aimed at getting all facilities to look for ways to prevent pollution and waste at the source. Starting with this simple premise, 3M claims that, since the program began, it has cut releases of polluting gases, fluids, sludges, and solid wastes in half, per unit of production. In plants all over the world, more than 2,500 projects have greatly reduced or eliminated pollution through development of new products, reformulation of existing products, and modification of manufacturing processes. Because of the program, 3M has won numerous environmental achievement awards, and many other companies have copied the program. 3P has also saved the company more than $530 million worldwide, and this amount is steadily growing because savings are calculated only for the first year of modified operations.

Some projects call for process redesign. At its videotape plant in Hutchinson, Minnesota, for example, 3M has added a "chemical refinery" to recover valuable solvents while dramatically cutting back on air pollution in the local community. Solvents used to help apply a magnetic layer to a plastic tape base are then evaporated and blown into tanks filled with activated carbon. This carbon absorbs the gas while exhausting clean air to the atmosphere. Steam then drives the solvents from the carbon so they can be distilled and prepared for reuse. The recovery system not only helps clean the environment, but also solves a critical supply problem for methyl ethyl ketone, saving the company some $5–7 million a year in chemicals that were formerly emitted into the air.

In Hilden, West Germany, a 3M plant producing decorative and reflective products had to dispose of 110 tons of cleaning solvents each year while spending 800 man-hours to clean storage tanks. Development of an automated, fully enclosed sedimentation cleaning and recovery system allows the solvent to be recovered from the cleaning solution while eliminating worker contact with the chemicals involved.

In Northridge, California, a 3M pharmaceuticals plant developed a water-based coating for medicine tablets that replaced a solvent-based material. The process modification required a $60,000 investment, but it also eliminated the need for additional pollution control equipment and cut 48,000 pounds of pollutants that would have otherwise been dumped into the atmosphere. At another 3M plant in Mexico City, a water-based resin replaced a solvent-based material for manufacturing cleaning pads, cutting down on air pollution in what is perhaps the world's most smog-plagued city.

Product redesign is sometimes necessary to overcome barriers to reuse or recycling. At a 3M plant in Aberdeen, South Dakota, manufacturing trim from face masks for respiratory protection had to be landfilled because the resin coating used to help the nonwoven masks hold their shape prevented reuse of the material. Brainstorming by a multidisciplinary team led to a modified design that eliminated the resin from the manufacturing process. The trim can now be reground and added to the materials used to make other nonwoven products.

Another five-year program at 3M called Challenge '95 involves the effectiveness of the company in using raw materials for its products. Any material that is not directly used, recycled, or converted to a by-product counts against production efficiency. One objective of the program is a 35-percent reduction in total waste. Even though the company estimates it has cut energy consumption per unit of output in half since 1973, it is seeking an additional 20-percent reduction in the Challenge '95 program by middecade.

Whenever possible, trim and other waste material is put right back into the production process. But in many cases this is not possible, so the company's Resource Recovery Department has cultivated widespread contacts, for both selling its waste materials and obtaining recycled products for its own use. Sometimes, even though recycled supplies are cheaper than virgin material, total costs can be higher because of greater process difficulties. 3M experienced this with a recycled resin it used in a product called Stak-It because of excessive lost production time. The solution was to have the recycled resin reprocessed to remove impurities. Now the total costs for using recycled resin match those for virgin materials. Waste plastic resins that have been resold by 3M are being used for a wide variety of products, including shrubbery pots, drainage tile, toy car tires, carpet fibers, plastic wood, swizzle sticks, fence insulators, office equipment, and many others.

Many factors in the design of products and processes determine recyclability of the finished products. Some considerations include using single materials for individual parts and pieces when possible and being careful about how joints are bonded together. Snap fits are easier to take apart than glued or ultrasonic bonded joints. Choosing recyclable materials, keeping careful records on which materials are used, and segregating scrap helps the process. Simple things like avoiding stuck-on labels can cut the costs of recycling.

The original 3P program started in 1975 was succeeded by a new program, 3P Plus, in 1987. Goals of the 3P Plus program include cutting all hazardous and nonhazardous releases into the land, water, and air by the year 2000 and cutting the generation of all wastes by 50 percent also by the year 2000. Air emissions alone will be cut by 70 percent through a worldwide 3M investment of $150 million in additional pollution control equipment by 1993. Even though many of these air pollution reductions will warrant credits that could be transferred to other 3M facilities or sold to other companies, 3M plans to return all the earned credits to local air quality control authorities, allowing a net reduction in total pollutants in the atmosphere.

Along with moves to reduce waste, the program also involves accelerated scientific research aimed at eliminating sources of pollution and boosting recovery and recycling of waste materials. Waste minimization teams have been set up in product operating divisions to find and implement opportunities for waste reduction at the source and for increased recycling. The teams include representatives from engineering, manufacturing, package engineering, marketing, research, and other appropriate units. An Environmental Leadership program linking the laboratories and marketing groups will use life-cycle analysis to address the key elements of product responsibility, from the conception and design phase, through manufacture, use by customers, and final disposition. The company is even offering a version of its famed Post-It notes made from recycled paper, including 10-percent postconsumer wastepaper, and after being used the notes still can be further recycled for use in such products as cereal boxes.

3M also works with customers to reduce their waste and pollution—including engineering assistance on process redesign—and offers to take back its packaging for recycling.

AT&T'S PROCESS IMPROVEMENTS

The changing industry viewpoint on the environment can also be seen in the semiconductor industry, where management's focus traditionally has been fixed on keeping on top of fast-changing technology. Electronic materials processing historically has been done rather poorly in the United States. The processes required are typically developed in laboratories and then simply scaled up for production in large plants, a method sometimes referred to as the "bigger beakers" approach. Advances in processing techniques have not been nearly as rapid as they have been in the chemical industry, where there is much more emphasis on process design. Since the final properties of the semiconductor devices being produced are so critical, industry emphasis has been on the quality of its final products rather than the processes used to achieve them. The view of management was that, since electronics was not like the nuclear industry or even the chemical business, environmental matters were of minor consequence. Any chemical processing required was seen as a small, benign activity.

The environmental movement, over time, has brought a major shift in attitudes in this industry, which can be illustrated by the experience of an AT&T semiconductor plant in western Massachusetts. Management had become aware of pollution problems at the plant, and progress was being made in cleaning up its processes. But in spite of these efforts, the AT&T factory was suddenly cited on the late news by a Boston TV station as being the biggest polluter in the Merrimack Valley. The plant came under additional criticism because of a couple of Superfund toxic waste sites that it was being forced to clean up.

Experiences like these brought a major shift in management's approach to environmental problems. Applying some total quality principles, such as those fostered by the Baldrige quality program, the company began to look for root causes. The hope was to get out of the reactive mode—trying to clean up smokestacks, discharge pipes, and hazardous waste landfills—and instead find ways to avoid creating pollution and waste in the first place.

AT&T looked to an industry ecology model for sustainable development, which expands the view of processing by starting at the mines where materials originate and following through to the ultimate disposal of all outputs. The ecological view of process design recognizes that there are limited resources as

INDUSTRIAL PERSPECTIVES AND INITIATIVES

well as a finite sink for waste products. An ideal solution, based on a systems view of industrial operations, is to create a total materials cycle, optimized to minimize residual waste and any effects on humans or the ecology.

One useful principle for industrial ecology, if it is feasible, is called "functional substitution." It is hard to plan for substitution opportunities, but when one does occur there can be very beneficial environmental effects. An example in the communications business is the substitution of glass optical fibers, made essentially from beach sand, for the copper previously used in transmission wire and cable. Industrial ecology involves far more than just technology. There are aspects of economics, education, political science, and the legal system that must all be considered if optimum solutions are to be found by decision makers.

In the case of semiconductor fabrication, AT&T's management hoped to find processes that were so profitable in themselves, either through avoiding disposal costs or by selling or reusing useful by-products, that there would be good reason to choose them beyond just being good corporate citizens. This could turn the quest for environmentally sound methods into an opportunity for the company, rather than following the old regime, which was based on just meeting applicable regulations so as not to be cited by the EPA. The company looked for turnkey systems that could accomplish these goals, incorporating process equipment or lines that might simply be installed without modification. Meeting such needs offers major opportunities to those in the construction industry. But until such turnkey facilities are available, electronics producers like this AT&T plant must work at modifying their own processes.

A good example of how this plant went about doing that involves arsine gas, one of the most troublesome materials required in semiconductor processing. A major use of arsine is as a dopant for the silicon used to make integrated circuits. Arsine is also the principal constituent of what are called 3-5 compounds (molecular structures combining atoms with three and five electrons in their outer electron shells), which are the basis of semiconductor laser diodes and a host of other high-speed electronic devices. In addition, arsine is extremely poisonous. If the state of California ever decided to go back to capital punishment using poison gas, arsine could prove much more effective than hydrogen cyanide. It's more poisonous than just about anything except phosphine, which is also used to make integrated circuits. The fact that it is so deadly makes shipping and stockpiling arsine extremely difficult. The question for the AT&T plant was, could a better way be found to make it, or could it even be eliminated? Considerable research was done trying to find other sources of volatile arsenic that are not as poisonous. Unfortunately, anything that was tried just didn't make acceptable semiconductors.

AT&T then took a step back. If the root cause of the material problem couldn't be completely eliminated, the next place to look was at shipping and storage problems. Could the arsine gas be produced right at the site and only when it was needed? Indeed, this approach did prove to be quite practical. The answer was to set up an electrochemical system using arsenic electrodes so that the arsine gas could be produced whenever it was required. The setup had to be designed carefully with attention to the types of electrodes and the use of proper voltages so that unwanted by-products were not generated, and so that the arsine produced was pure and in the right quantities. The process has progressed from the laboratory to pilot lines in the factory so that it can be used on a trial basis. A manufacturer in New Jersey is now planning to sell these arsine generators to other semiconductor makers. The chemists who developed the process have gotten accolades from coworkers, friends, and families and have gained considerable satisfaction from developing something recognized by the public as a positive contribution for the environment.

Another troublesome area for electronics manufacturers is the use of chlorofluorocarbon (CFC) compounds as solvents for cleaning circuits, particularly before and after soldering. The Montreal protocols called for reduction in the use of CFCs because of worldwide agreement that they are linked to the depletion of the ozone layer at the top of the atmosphere that filters out harmful ultraviolet radiation from the sun. The chief executive of AT&T's Merrimack Valley facility promised that the plant would phase out the use of CFC solvents according to the international agreements. His promise, however, was driven more by input from lawyers and public relations specialists than from technical considerations. Fortunately, although the task has proven to be difficult, the technical people at the plant have been able to find ways to live up to the promise. In fact, the plant has kept well ahead of the phased reductions called for in the Montreal protocols, which is fortuitous because new findings about the fast-spreading hole in the ozone layer have induced policy makers to shorten compliance deadlines even further.

The answer to replacing CFCs was found in solvents called terpenes. These naturally occurring compounds are found in the essential oils of many plants and are thus environmentally benign. But in these applications, the terpenes are used in closed systems along with organic materials. Since some of the organics are flammable, new cleaning systems had to be devised. This required a shift to water-soluble soldering fluxes. The firm's scientists are now examining what seems like a mundane area—the physical metallurgy of soldering along with the kinds of fluxes used and how they are applied. That's because revising such processes often forces a holistic approach, going back and reexamining fundamentals in order to find more environmentally sound alternatives. There are no magic formulas. The efforts of this pathfinding AT&T plant have not gone unrewarded. In 1992, it was a winner of a Malcolm Baldrige National Quality Award for total quality.

Over the coming decade, expected advances in semiconductor technology will result in the need for new types of complex silicon processing lines that may cost from $700 million to $1 billion each to build. Many steps are required to produce integrated circuitry, including growing crystal, applying photosensitive layers to tiny flat chips of crystalline material, coating the surfaces, and then exposing them to light through photolithographic masks so they can be selectively etched. Different portions must be doped with chemicals such as arsenic and phosphorous as the process proceeds. As feature sizes of the circuit elements decrease to less than 1 micron, contamination of the surfaces by minute particles of dust or lint could seriously damage the circuitry, putting extreme demands on a clean-room environment. As a result, new processes might be needed, such as molecular-beam epitaxy, that can operate completely within a vacuum.

Because of the changed views in the electronics industry, the semiconductor factory of the future is likely to produce benign products using processes that avoid any environmentally harmful side effects. The future electronics plant will be highly energy-efficient, producing some useful by-products but with a minimum of residual waste and very few defective devices. Process lines will be built from modular sections, and cluster processing will enable most operations to be performed within vacuum chambers without any need to remove partially completed devices into the surrounding clean room. Designers of process lines will aim for turnkey operation for clients, yet designs will be geared to reuse. The lines will be flexible rather than monolithic, allowing producers to fabricate short runs of items for particular customers easily and economically, rather than requiring runs of a million or more of the same devices in order to be cost-effective. Stockpiling will be unnecessary since these processes will allow circuitry to be fabricated to order.

POLLUTION REDUCTION IN ENERGY PRODUCTION

Utilities and other heavy users of carbon-based fuels, such as petroleum refiners and metals processors, also have an excellent opportunity to reduce pollution resulting from their processes. The original Clean Air Act (CAA) of 1963 concentrated on cutting down on local emissions problems, particularly by controlling ground-level concentrations of nitrogen and sulfur oxides within 50 miles of discharge. This allowed plant operators to meet regulations partly by building smokestacks higher. The result was an increase in the amount of pollutants that are carried long distances in the atmosphere.

The Clean Air Act of 1970 mandated stringent emission standards for any new large industrial plants or electric utilities emitting sulfur dioxide. But the older plants did not have to meet these strict requirements. As a result, the Project 88 Report estimates that by 1995 over 90% of utility discharges of sulfur oxide compounds will come from older plants. Some of the shortcomings of the

POLLUTION REDUCTION IN ENERGY PRODUCTION

original legislation were addressed in the 1990 amendments to the Clean Air Act. Higher smokestacks, for example, can no longer be used as a means of avoiding ambient restrictions. Still, electric generating plants remain a particularly opportune target for both greater energy efficiency and reduced pollution.

Unfortunately, in the United States, since the mid-1970s when conservation of electricity became a major element of public policy and regulation, the focus has been almost completely on the demand or user side of the electric meter. Even though great advances have been made in technology for electric generation, the replacement of older plants has slowed to a crawl. New equipment and systems offering both greatly improved thermal efficiencies and significantly reduced environmental emissions include gas turbines, used both singly and in combined cycles; the integrated gasifier combined cycle; new humid air turbines; and several others. Yet, because of the emphasis on conservation and the economic troubles of many utilities, there have been few new supply-side additions or replacements in recent years. Because of the perceived advantages in the short term of reduced capital expenditures, most utilities have strong programs to extend the lifetimes of existing generation systems.

The net result in the United States has been the aging of the electric-generation infrastructure as capital spending has been cut back, along with retarding R&D on improved generating systems, while almost destroying electric-power-equipment manufacturing capabilities. These policies have directly and significantly added to the environmental pollution produced by the electric power industry.

Neither policy makers nor electric industry executives have paid much heed to analyses suggesting the folly of the present course being followed in the United States. Recently, instead of the top-down approach favored by economists, the MIT Energy Laboratory did a bottom-up, engineering-based study of New England power systems that not only verified the economic analysis, but also demonstrated that a more energy-efficient, less-polluting route to the future would likely be accompanied by cost savings. Multiple simulations were conducted of future configurations of the New England electric system over the next twenty years, including variations in end-use conservation, new supply technologies, life extension of existing plants, load growth, and fuel prices. Results showed clearly that energy efficiency of both supply and demand will produce the most economic electric power over the period while substantially improving the environment.

The Energy Lab analysis showed that new generating technology can reduce emissions of sulfur and nitrogen compounds as well as carbon dioxide with very small changes in the cost of electric service. The higher efficiencies of the new generators would reduce the fuel required per kilowatt generated, and fuel costs are the dominant factor in operating costs. Cost reductions due to fuel savings would virtually offset the increased capital charges for the new or modified plants with properly configured dispatch logic for the New England power pool. Although the results of the New England case cannot be transferred automatically to other regions and countries, economic dispatch of power systems is common around the world. Therefore, findings such as these are pertinent for electric power planning, and for environmental improvement, in any region.

Another important area for consideration by policy makers, according to the Project 88 Report, is that the market sets prices for different types of coal, oil, and natural gas without reference to the environmental consequences—and costs to society—of their extraction and use. The Public Utility Regulatory Policies Act (PURPA) of 1978 was written to encourage the use of alternative sources of energy when oil and gas prices were rising rapidly. Because of changes in the subsequent energy markets, particularly a glut of world oil supplies and falling prices, the intended incentives for alternate energy sources did not materialize. Policy makers need to find ways to factor in the full costs of alternate energy sources so that industry and utilities are encouraged to make environmentally sound choices.

EMERGENCE OF THE ERA OF "SUSTAINABLE DEVELOPMENT"

There are signs that, just as an era of environmental neglect by industry was replaced by one of begrudging compliance, a new era is beginning. The changes it will bring are so dramatic that it is being seen by many as a shift to a new environmental paradigm. It might be termed the "era of sustainable development."

Recognition is growing among business executives that continued destruction of the environment is a no-win situation. Unless there is progress toward a cleaner environment, citizens' groups and the political process will steadily tighten the reins on industry's activities. Smog-laden air and polluted waters effect all citizens, including business executives and employees, and industry and its products are increasingly being held accountable for environmental decline. Decision makers are finding it harder and harder to ignore the rising clamor for reform. Continuing business as usual will therefore mean that both the public and industry will be losers in the long run.

This concern is leading to the concept of sustainable development. This new view requires that environmental and economic factors be considered jointly in planning for the future, so that development is not eventually stifled. Ways must be found to limit the generation of waste and pollution while better controlling or disposing of any waste products that are created. Vital parts of the ecosystem also need to be preserved or replaced, such as greenery to convert carbon dioxide to oxygen and the gene pool provided by a diversity of species around the world. Otherwise, public opposition to industry's practices is likely to intensify to the point where either new facilities cannot be built, or, if they are built, the costs of compliance would be so high as to make them uneconomical. This has already happened in the case of nuclear power plants in the United States.

But the emerging paradigm goes even further, to a recognition that environmentally sound practices and processes may actually prove more cost-effective than existing methods, as 3M and AT&T are discovering. In some cases, not only have process changes eliminated wastes, they have also proven to be more efficient. Sometimes this is accomplished by cleaning and reusing costly materials, such as solvents. In other cases, alternate uses are found for by-products that were previously discarded as waste. Barriers to reusing or recycling materials can sometimes be eliminated by redesigning products, changing materials, or altering chemicals used in processing.

Underlying the new thinking is the belief that ways can be found to make development compatible with environmental management. Reaching a state in which economic growth becomes environmentally benign, however, will require major changes throughout both the industrialized and the developing world. Integral to achieving sustainable development is the need for users and producers to begin to internalize the long-term environmental costs of their decisions and choices.

THE NEW VIEW: ASSET MANAGEMENT

A new perspective is required recognizing that each nation is endowed with a unique set of national assets. These may include natural resources; human and financial capital; and a wide range of inputs to productive capacity, such as an electric power grid, highways, and telecommunications networks. To some degree at least, each nation has some control over the deployment of these assets. An asset-management perspective requires each country to consider *both* the preservation and the expansion of its total assets. The old concept of "growth," which generally drove national efforts, needs to be replaced by a new view of "development." Growth implies an expansion in size with minimal changes in the characteristics of the economy and the social order. Development, by contrast, must be viewed as a transformation of both the economic and social system to a more effective framework, with a shift in emphasis from the issue of scale to that of composition.

The new paradigm is still evolving, and widespread debate continues on its implications—not just for today's technologies and industrial processes but also for issues such as societal organization. International mechanisms are being sought to address environmental problems such as preserving of the rain forests, controlling population growth, and curbing the generation of greenhouse gases. Economic issues are also deeply entangled in the debate, as the world community struggles with the question of how the more advanced states can best help in solving the rapidly escalating environmental problems of the developing world and the former Eastern bloc nations.

Since each nation has a unique set of assets, each will need to determine its own route to a sustainable-development culture. There are worldwide issues, however, that will have to be dealt with by the international community. Already progress is being made in controlling global pollutants, such as the CFCs responsible for the growing holes in the ozone layer. There are also efforts by industrialized nations to bring the latest technologies to bear on the escalating pollution and waste problems of the developing world. In the future it may be necessary to find ways to control the movement of industry from areas with more stringent environmental requirements to those that are more lax.

Nations trying to cope with growing environmental problems must find ways to adjust traditional economic systems so that they take account of the costs of environmental degradation. Each country may need to develop appropriate policies so that producers and consumers will automatically be induced to internalize environmental costs. National policies also need to encourage rapid development of appropriate technologies to help each separate region to achieve the sustainable-development ideal. An encouraging start has already been made on addressing such problems. Dozens of nations summarized the major environmental problems that each of them faces as part of their contribution to the Earth Summit held in Rio de Janeiro in June 1992. Developing policies and mechanisms to address the myriad problems identified in these reports will undoubtedly require diligent international efforts that must continue for decades to come.

EXPERIENCE IN EUROPE

Many European firms are faced with situations similar to those in the United States, but they take a different approach to policies and regulations which, as will be seen in Chapter 5, does not lead to as much litigation as we have in the United States. This means that more new ideas can be tested more rapidly without the fear of litigation. For example, a number of elements within an integrated-systems approach are suggested by Janusz Niemczynowicz of the University of Lund in Sweden. He recommends small-scale projects using multidisciplinary cooperation rather than "technological monumentalism." He recommends that source control and pollution prevention must be encouraged over "end-of-pipe" pollution control, and local disposal or reuse are preferable to transporting waste. In Europe, these principles are already more accepted than they are here in the United States. For instance, treated wastewater is used extensively for irrigation and industrial processes. Ash from incineration can be used in road construction, and sewage sludge mixed with organic solid waste can be composted. Wastewater treatment plants using new biological processing systems may prove far cheaper than traditional facilities, Niemczynowicz suggests. Existing ecosystems, such as wetlands and riparian zones with root uptakes, can be successful in treating waste and providing habitat and other environmental benefits. According to this philosophy, holistic approaches, based on the latest scientific knowledge and modern technology, must be applied in dealing with the growing environmental threats posed by rapidly growing urban areas in developing nations and by the problems of Eastern Europe.

Swedish specialists are applying these principles in helping Polish officials address serious pollution problems in the Zabrze region of Poland, where health statistics show the devastating impact of intensive industrialization without environmental controls. Life expectancy in Zabrze is seven years shorter than it is in Sweden, and the death rate for infants below one year of age is five times higher (compared to three times higher in all of Poland). Solutions proposed by the Swedes include construction of an efficient centralized heating plant which will eliminate the pollution from less efficient coal stoves currently used in homes. They have also proposed the installation of an industrial pretreatment plant that can process river water to standards pure enough for industrial use. This will free up badly needed drinking water for the inhabitants. Small lakes will be created, and the river and its tributaries will be forced to meander where they are now straight, creating filter strips of vegetation and seasonally flooded riparian zones to increase natural purification. Industrial processes will be subject to increased maintenance to stem the flow of pollutants into the environment from easily preventable sources such as leaky pipes. The integrated approach, being carried out collaboratively by technical specialists such as hydrologists and biologists, along with city planners and politicians, is expected to cost about a fifth as much as a large treatment plant for river water than had originally been planned.

Collaborations among enterprises to achieve an "industrial ecosystem" is another concept being explored in Scandinavia. At Kalundborg, Denmark, just outside Copenhagen, an electric generating facility, a cement production firm, an oil refinery, a plasterboard factory, a sulfuric acid company, and a biotechnology plant all work together to achieve a prosperous "ecology/economy." Each of these firms use the other's by-products, creating a closed-loop system where "waste" is continuously recycled.

Knowledge gained from environmental projects such as these should help U.S. policy makers to find better ways to stimulate progress in this country. Policies will emerge that continue to foster the development of the thriving environmental industry, offering a steadily improving set of cost-effective options to those faced with source prevention, pollution, or treatment problems. The construction industry should play an increasingly important role in each of these areas to ensure that the programs are most effective and efficient in implementation.

In conclusion, it is clear that the attitudes of industry toward environmental protection have been steadily evolving, from neglect in the early part of this century; to begrudging regulatory compliance; and more recently, in some cases, to recognition that investing to eliminate pollution and waste can actually be profitable. This latest transition still remains in its early phase. While changing attitudes were once driven by the idea that we are approaching the limits of critical resources and the earth's ability to handle waste, there has been a shift more recently to the notion of sustainable development. Instead of just trying to manage waste and pollution, this viewpoint shifts the focus of environmental protection toward source control, when it proves feasible, to prevent the generation of waste and pollution in the first place. It also encourages policy makers to try to incorporate more rewards and penalties right into the market system to encourage environmentally sound behavior. Even though a start has been made on international collaboration to attack global environmental problems, it is proceeding with recognition that each nation faces its own unique set of issues and that solutions must be developed within the social and political context of each individual region. In the next chapter, we will explore the evolution of environmental policy making and regulatory approaches in more detail.

CHAPTER 3

ENVIRONMENTAL POLICY AND REGULATIONS

This chapter examines the emergence of the environmental movement, the processes that propel policy and regulation development, along with the current directions of environmental policy and the diverse approaches to regulation growing out of them. It is not our intent to provide an in-depth look at the policy making and regulatory directions in every different market segment. Rather our objective is to impart a general understanding of the process and problems that face policy makers and regulators and to review some of the history and current trends in these areas. We have not specifically distinguished policy making and regulation as separate topics, but present them together in a discussion of the basic regulatory approaches in a few domestic market categories. The chapter concludes with a look at some European views on how to balance industrial development and environmental preservation.

As is true for many public policy areas, solving environmental problems usually necessitates cooperative action on the part of a variety of actors in society who are generally not affected equally by the outcome. Not only may the results of policies have an unequal impact on various parties; in many cases, disputes arise because different groups have directly opposing interests. Such conflicts must be resolved or mediated, sometimes through compromises that satisfy neither side, before cooperation can be achieved. In most cases, organizing effective, cooperative response requires society to call on government to consider solutions, direct the implementation of programs, audit and coerce compliance, manage conflicts, provide information and education, and perform a variety of other functions that support the resolution of the environmental problem. Therefore, the government, as policy maker and regulator at every level, plays what is perhaps the most significant role in the environmental market. In many ways, the business climate in environmental markets is not autonomous. Rather than being driven by traditional forces of supply and demand, market directions are more likely to be determined by a complex set of government actions.

Participants in environmental markets must understand the role of government, as well as the directions in which policy makers wish to drive the system. This will help them make appropriate business and strategy decisions so that their organizations can adapt to changing markets and tap new market opportunities.

Change seldom comes without controversy—but at the same time it is controversy that induces change. Over the past few decades, great alterations in U.S. society have come about through concerted action against established practices, including protest marches and occasional civil disobedience by a wide range of organizations with specific agendas. Among the successes achieved by such groups are progress in civil rights for minorities and women, the government's withdrawal from the Vietnam War, and changes in children's programming for television. Citizen-initiated litigation has also grown to support social causes. Through such activities legal precedents have emerged for citizen action. Grassroots groups have become important and powerful political influences given their significant fund-raising skills, polished media campaigns, and channels of access to legislators. Nongovernment organizations (NGOs) have become more important in driving policy since there has also been growing distrust of government and the bureaucracy. The rise of such groups has been fueled by the widespread belief that some officials are more willing to cater to powerful interests than to try to protect the general populace. Increasingly, within this combative setting, the focus of public attention has been turning to environmental matters.

The rise of concerted public action has forced policy makers to operate within an increasingly turbulent and difficult social context. In some cases, goals that may seem worthwhile for the public as a whole have been achieved through organized citizen action, such as the cleanup at Love Canal. But this may not always be the case. The existence of powerful citizens' groups with narrow interests may increase the potential for some of them to use high-pressure tactics and legal precedents to obstruct or counter actions that also may seem to offer aggregate benefits to the society. For example, the "not in my backyard" or NIMBY syndrome makes it increasingly difficult to find sites for landfills and solid waste combustion plants. This is not to say that sometimes such obstruction is not justified, perhaps because of a threat of toxic pollutants from inadequate controls. But the success of such actions may thwart efforts to develop suitable options for managing waste. As environmental and other types of groups gain power and influence, they can target agencies and obstruct attempts to find new solutions to environmental problems, just as in past times industry opposed efforts to stiffen environmental rules.

An underlying premise of the American system is that out of the open clash of opposing interests within a structured process of decision making, balanced solutions will eventually emerge. But the reality may be that, as one set of proponents gains influence over others, or compromises reshape decisions to satisfy conflicting interests, the solutions reached may be less than optimum for the society as a whole. Finding optimum solutions aimed at producing the maximum environmental improvement with the least disruption to the market system and society, often involves a delicate balancing act even when decisions are carefully based on factual data and experience in the field. But political pressures, whether exerted by industry trying to reduce its costs or by local groups opposed to any waste disposal facilities no matter how sound the design, can badly distort the outcome of the decision-making process. For example, in the battle over tradable permits under clean air legislation, legislators from states that produce large amounts of high-sulfur coal tried (unsuccessfully) to forbid fuel switching and instead force more costly flue-gas scrubbing. This would have raised costs for midwestern utilities, triggering political maneuvering for nationwide cost sharing. Such an approach would have distorted the fundamental "polluter pays" principle underlying incentive programs. In the waste-management industry, Congress is now struggling with the interstate transportation of waste. Although many complex issues are involved, part of the dispute over what if any limits should be applied revolves around industry's contention that it is often less costly to haul waste to remote locations. Some environmentalists and regional groups insist, however, that disposal should be limited to the localities where the waste is generated regardless of the costs. Too often clashes such as this result in one side or the other gaining a victory, with neither side being willing to consider alternatives that would both satisfy public needs and minimize overall costs.

Even without intense battles between groups supporting conflicting causes, the environmental arena often presents multifaceted sociopolitical and economic problems not amenable to easy solutions. A holistic view may reveal complex policy tradeoffs that are not readily apparent. For example, more recycling of newspapers in the .nited States appears to be a logical, environmentally sound policy objective. Yet a complete life-cycle analysis might show that undesirable increases in other types of pollution might be the end result. The case for recycling is supported by landfill studies which show that the fastest growing segment of the solid-waste stream in recent times has been paper, largely newspapers. Yet paper recycling in or near metropolitan areas requires electricity generated by burning fossil fuels, thus releasing greenhouse gases into the atmosphere. By contrast, most virgin newsprint comes from Canadian regions served primarily by hydropower. Perhaps policies encouraging careful forest management in the northlands while shipping most U.S. newspaper waste to parts of the world where paper is scarce and costly would offer an alternative worthy of consideration. This might relieve pressures on overburdened landfills in the United States while reducing greenhouse gas emissions and also helping to preserve threatened woodlands in other parts of the world.

Modern social and economic practices are, in many cases, being recognized as destructive exploitation of the environment. Intensive agriculture, for example, may waste precious water, deplete soils, and foul waterways and groundwater because of the cumulative runoff of chemicals and suspended solids. Resource extraction, including strip mining, conventional mineral and metal mining, and clear-cutting of forests, can leave entire regions devastated. Public disapproval of mounting environmental problems such as these is forcing a reexamination of traditional methods and practices. Questions are being raised about basic production processes, consumption patterns, and the causes of excessive generation of waste and pollution, as well as the types of policies governments might employ to foster more environmentally sound approaches without causing severe economic disruption. This broad public concern is creating a demand for new solutions that reduce or eliminate wastes. Sometimes this requires extensive redesign of the basic means of production, of products themselves, or of associated factors, such as packaging. Since all waste cannot be eliminated, better methods of treatment or reuse are being sought, or alternate markets are being found for the polluting by-products of industrial processes. In some cases, product redesigns and process modifications have not only resulted in less environmental degradation, they have actually become a new source of profits.

Until recent times, it was widely believed, especially within industry, that it was impossible to preserve environmental values while allowing development. Analysis of past data indicated a positive linear relationship between the growth of GNP and energy production, a major source of atmospheric and other forms of pollution. If an economy grew, then it required a commensurate growth in the energy generated, predominantly electricity in industrialized societies. A similar positive linear relationship was believed to exist between the amount of electric power generated and the pollutants emitted by a combination of electric utilities and other industries. This led to the conclusion that, as an economy grows, the consequent pollution will increase in direct proportion to the growth rate. In fact, the belief in this relationship was so compelling that in industrial regions the amount of fumes being emitted by billowing smokestacks once was considered a fairly accurate gauge of economic prosperity.

Now society recognizes that such relationships are not inevitable. Actions are possible which lead to a decoupling of causes from what had been assumed to be unavoidable effects. Energy conservation and more efficient systems for generating, storing, and transmitting energy, for example, can permit economic growth without a burgeoning need for electricity, and cleaner methods for producing energy and treating effluents can cut the levels of pollution associated with a given increase in development. In spite of steady economic expansion in the United States, there are a number of once heavily polluted rivers that have recently seen a revival of fish populations. This is the case for the Willamette River in Oregon, for example, where reducing pollution flowing from industries located along the river has helped to reestablish fish stocks. Similarly, Atlantic salmon have been reintroduced into some Maine rivers in recent years as pollution has been cut back through tougher regulation and enforcement.

POLICY PERSPECTIVES AND THE ENVIRONMENT

It is not only in the highly developed countries such as the United States that there has been an unprecedented rise in organized concern for the environment. Green movements have developed in every country in the world, regardless of political systems, economic status, or prevailing religious philosophy. To probe the future of the movement, it is necessary to ask about its past—why does environmentalism grow everywhere, and what common trait or traits promote it? Is it an innate human response, or could it be a natural outcome of public order as asserted by governments?

The mutually shared characteristic that leads to environmental consciousness is simple, but profound. Growth of what may be labeled an *environmental constituency* is the vital link that joins environmentalists of every nation. Environmental constituencies are groups that share common values and feel a common plight. These groups do not necessarily exist within a single geographic region, but they do share knowledge about a particular problem and often agree on how to resolve it. Concern for the welfare of some aspect of the environment is the fundamental reason for the formation of the constituency. In most cases participating in a group of those with similar opinions is the reason for joining in the activity. Environmental groups cannot be viewed as simply another trend that will pass when public attention wanes. In fact the green movement has been gaining more and more followers in all regions of the world and the range of issues involved has been widening as well. These trends are becoming increasingly evident at the international level as acceptable solutions to environmental problems in one geographic area are encouraging similar approaches in other regions. Environmental values will gradually become ingrained into all activities undertaken by society as the will to guard the earth's resources spreads around the globe.

Many environmental efforts initially take root at the local level and then spread to national proportions, particularly in highly industrialized, democratic nations. The growing influence of the media in recent times, especially television, has helped to educate millions of citizens in many nations about environmental issues. This increased access to the latest information serves to broaden environmental awareness beyond local problems, such as smokestack emissions and discharges into waterways—to worldwide concerns, such as threats from global warming, acid rain, and the depletion of the ozone layer. In the United States alone, membership in environmental organizations is reported to have grown from 3 million in 1970 to more than 20 million today, and it is still growing. The growth in groups that can put pressure on the government has been paralleled by commensurate increases in policy making and regulation.

Following through a simplified scenario of how new environmental initiatives get started provides insight into the process of environmental management. Staying alert to directions and trends developing in this way within the public arena can help companies in forecasting future market trends. The process begins when environmental constituencies, backed by science, economics, and political power, pressure policy makers for appropriate changes in the government's policy. Policy makers weigh gains and losses in the political arena, addressing problems that are most strongly advocated by large constituencies first. Law and regulations ensue from this process. Therefore, new market segments take shape as concerns from environmental constituencies pass through government's policy-making machinery and emerge as law and regulations. These then provide business opportunities, such as for hazardous waste cleanups, asbestos abatement, and lead paint removal.

Managers must understand the policy implementation process and be aware of what outcomes to expect from it so that future opportunities will not be missed. The construction industry can even participate in the process, encouraging environmental programs that are not just worthwhile but also represent business opportunities. Environmental law and regulation must be studied and understood since these are the mechanisms that actually establish and shape various market segments.

Currently, the markets for environmental services are also changing. The trend toward source control as a principal means of waste management, rather than the traditional approach of simply adding waste treatment mechanisms, has become a worldwide phenomenon. Advocates of this change, part of the larger environmental paradigm shift known as sustainable development, focus on the elimination of waste-producing activities through modifications in the existing means of production, including materials, technology, and management methods. This philosophy signals new directions for the construction industry and offers emerging opportunities for innovative firms to participate in the environmental movement. In the future, constructors will not only be responsible for building and implementing the technologies intended to control pollutants, but will themselves promote sustainable growth by directly attacking the causes of pollution and waste at the source rather than simply limiting potential environmental damage.

ENVIRONMENTAL POLICY: HOW AND WHY

Environmental policy is determined through a variety of considerations (what academics may refer to as policy perspectives) that include science, economics, law, public administration needs, and environmental ethics. A wide range of regulatory approaches have been tried by policy makers to satisfy contending demands and to address the often unique problems within different environmental sectors. Dealing with the careless disposal of batteries containing toxic materials by consumers, for example, requires quite different policies from those aimed at getting chemical plants to reduce emissions of harmful gases. Some of the major forms of regulation, along with their strengths and weaknesses, will be reviewed here. What is clear is that no one approach is always best, so that a variety of regulatory schemes will continue to be needed depending on the demands of each situation. It is also clear, however, that there is a continuing evolution in the underlying approach to regulation. Fundamental to current regulatory directions is the emerging view that, if economic growth is to continue, development must be *sustainable*—otherwise, either growth must eventually stop or its effects will stress the environment past tolerable limits for sustaining health and even life.

Because of the wide range and complexity of issues across the total marketplace, firms should not attempt the impossible task of monitoring the entire environmental industry. Instead, individual companies choosing to do business within a particular niche should concentrate only on selected market segments pertinent to their ventures.

The bulk of environmental policy in the U.S. domestic market can be divided into six general categories: air pollution, water pollution, solid waste, hazardous waste, toxic substances, and nuclear waste. Each category might be further subdivided into any number of market segments based on type of service, customer type, technology, or geography. Overall policy trends are important for a general understanding of the nature of the industry; but if firms plan to participate in any particular segment of the market, they should consider related policy making in that area as essential knowledge.

DEVELOPMENT OF ENVIRONMENTAL REGULATION IN THE UNITED STATES

A brief review of the history of environmental regulation in the United States demonstrates that broad policy changes have affected the emphasis of regulators and lawmakers over time. Although generalizations are not completely applicable across all areas of policy making and regulation, some trends have emerged.

Until the Johnson era in the 1960s, government regulation, especially at the federal level, was centered on adjusting imbalances in "traditional" economic factors. Policy makers were primarily concerned with controlling the influence that firms could exert in the marketplace. Early regulation managed the pricing practices (among other competitive traits) of railroads, oil companies, pipelines, radio, and public utilities such as interstate electricity. During the New Deal era beginning in the 1930s, federal regulation reached financial markets, trucking, communications, and airlines. However, none of this federal influence was directed outside the arena of traditional economics and markets. In the 1960s and 1970s, however, federal policy makers and regulators undertook a radical departure from what had been common practice up to that time: *social* regulations emerged. This was a major shift toward regulating the cumulative affects on the overall society of the actions of a broad range of industries and even government entities, such as sewage treatment plants. The change toward placing greater curbs on the behavior of industry and municipalities coincided with the shift in attitudes about the capacity of ecosystems to absorb ever-increasing amounts of waste and pollution.

During the Johnson era and beyond, the American economy became subject to an increasing array of these new types of regulations. The change was characterized by the creation of legislation, regulations, and new agencies organized along functional or issue lines rather than by industry categories. The result was the development of sweeping powers housed in functional agencies that had jurisdiction over some particular aspect of "all business activity." One of the first steps in this direction was the Air Pollution Control Act of 1962. Other social regulation included the Food, Drug and Cosmetics Act of 1962, The National Environmental Policy Act (NEPA) of 1969, and The Occupational Safety and Health Act (OSHA) of 1970. This spurt of newly directed social regulation did *not* die out after the turbulent 1960s, however. In fact, after 1970 the pace of social regulation actually steadily increased.

Environmentalism was an important driving force in the move toward increased social regulation. The establishment of the EPA was a significant step in the consolidation of the environmental regulatory movement in the United States. In December 1970, this new agency was officially created and would bring together a variety of previously disparate functional groups under a single governmental structure. Past environmental enforcement had been largely ineffective because earlier statutes included unwieldy provisions for bureaucratic blocks in the enforcement process. Because of such flaws, targeted standards remained unrealized. Also, the regulated community was often successful in promoting unenforceable standards and practices, which added to the impotence of the regulations that did exist. As the consolidation took shape and the agency's power grew, Congress enacted stronger environmental laws, and the EPA became more adept at enforcing them.

COMMAND-AND-CONTROL REGULATION

Whereas it was fairly straightforward to regulate particular industry or market sectors—by setting rate structures for shipping goods, for example—it proved much more difficult to find ways to induce a broad range of industries to modify behavior so as to accomplish goals such as improved air quality and cleaner rivers and streams. Early environmental regulatory emphasis centered on what was called "command-and-control regulation," which is a policy of pollution control based on establishing *standards*. These standards might limit a concentration of a particular pollutant in a receptor, regulate a particular emission level, or require implementation of a particular technology. Standards were usually loosely based on nonquantitative goals, such as "water suitable for drinking or swimming," which were set by policy makers and members of the affected community. Regulators embraced standards because they provided concrete goals as well as measuring sticks for gauging effectiveness. It proved remarkably difficult, however, to translate these nontechnical desires and goals into actual enforceable standards.

Standards that limit the concentration of a particular pollutant at a certain receptor, also called ambient standards, were among the first promulgated by government. These were intended to limit the concentrations of certain pollutants in particular media (air, water, soil, and so on). They were a logical first method of regulation since scientists generally thought that some small concentration of pollution was acceptable in most cases. These laws were intended to keep the pollution levels below some set of tolerances.

Regulators used ambient standards as aids in determining other types of regulatory measures, such as emission and technology standards. For instance, New York City once did an analysis that led regulators to set an effluent standard for sulfur dioxide using an ambient standard of 0.02 parts per million (ppm). Unfortunately, the process employed to come up with this limit turned out to be flawed. After determining the actual ambient concentration of the chemical and its sources in the city, analysts determined what the maximum permissible output of each source would have to be in order to maintain that level. Many factors were not considered, however, including meteorological conditions, the ability of sources to introduce or disperse pollution, the geographic distribution of polluters, and the future growth in pollution due to industrial expansion. As a result, it soon became clear that the overall goal could not be achieved using the source limit that had been chosen.

Emission or effluent standards, which dictate specific levels of pollution that a particular source can emit, have their own special problems in addition to the difficulties described above. The process of translating overall criteria into specific performance levels was often not performed, so that regulators failed to come up with clear-cut requirements. Instead, regulations would call for "good practice" or "best available technology." Worse, large and powerful business interests that were affected by the regulations often entered into protracted and costly litigation that slowed both implementation and enforcement. The command-and-control process also implicitly required regulators to catalog, assess, and monitor millions of businesses to ensure that they installed the proper equipment. Many regulators continued to advocate this method of pollution regulation because they believed that compliance could be simply determined as long as some sort of monitoring or measurement system was installed.

Difficulties associated with this approach have become more apparent as the need to control new pollutants has arisen. Many of these newly regulated substances cannot be detected through visual inspection, smell, or any other simple means, so that detecting violators has become increasing difficult. While sulfurous fumes and particulates belching from a smokestack may easily be seen and smelled even without monitoring instruments, carbon monoxide and many other organic compounds are much more difficult to detect and measure. The fact that many of the worst, largest, and most visible polluters have cleaned up their large plants and processing complexes only makes the job more difficult for regulators. As a result, officials must now deal with many more firms at widely scattered locations to make commensurate gains in pollution control.

Technology-specific standards require a firm, set of firms, industry, or group of businesses within a geographic area to install some device that controls emissions. While the concept may seem simple, it suffers from many of the same drawbacks as emissions standards—including the need to set the standard, the potentially large number of firms in the regulated community, and the tendency of large polluters to fight implementation. Technology advances may also be stifled by technology standards since firms have few incentives to overachieve once a standard is set.

Another type of technology-specific standard is the requirement of "best available technology." This, too, provides disincentives for technology development, since firms that need the technology—and are best positioned to develop it—will avoid doing so because of the fear that regulators would then force them to install it. Furthermore, if regulators do force industries to continually upgrade facilities, costs might become prohibitive, especially since most capital pollution-control improvements require substantial investment. This might force many firms out of business. While loss of some businesses is inevitable, most regulators recognize that it is unacceptable to impact a market too severely.

Critics of command-and-control policies also charge that the policies do not provide the most environmental benefits at the least cost. Even though the approach might achieve the aggregate level of environmental protection desired, the critics contend that the total cost to society is much greater in terms of compliance than alternative approaches. One reason is that the costs of meeting standards vary widely, not only among industries, but also among firms competing in the same industry, and even among those in the same geographic location. For example, should a large sewage treatment plant that discharges effluent well out to sea meet the same standard as a nearby factory discharging into a small stream? Also, large firms that may be able to meet standards at a very low cost per unit of pollutant control often have an advantage over smaller plants, which must spend proportionately much more on compliance. Studies have suggested that ratios of 10 to 1 in control costs per unit are not unusual; and, in extreme cases, differences could range to 100 to 1 or more, depending on the age and location of plants and the available control technologies.

Since real-world investigation suggests that uniform standards could prove quite unfair in some cases, it might be worthwhile for the government to seek a practical system for imposing nonuniform standards. These would achieve a certain amount of pollution reduction in the aggregate without specifying particular actions for individual regulated firms. For example, instead of dictating rigid percentages, the regulated community might reallocate reductions among themselves to equalize the cost per unit of pollution reduction.

It is clear that nonuniform standards would fail if dictated by government alone. First, government is not privy to the information needed to perform the analysis necessary to set nonuniform standards because the analysis involves details of the operations of perhaps many thousands of industrial plants. If forced to provide cost data themselves, firms would have the incentive to exaggerate or hide information—even to lie—and it would be difficult or impossible for regulators to verify or dispute many of these claims. Moreover, the scheme might never be realized because, as soon as different standards for different firms were proposed, there would immediately be an intense debate over its fairness. The pressure from accusations of favoritism could lead to time-consuming delays or outright failure.

Nonetheless, it does make good sense to try to achieve a target level or overall aggregate cap on emissions by somehow reallocating the pollution-control burden using a scheme to allow polluters themselves to determine how to meet the challenge. Those who could control cheaply would take on an added share; and those finding it exceedingly costly would, in turn, share some of the burden. This is not such an unusual approach. It is the basis for the functioning of all market-based economies—allocation of goods by buyers ands sellers in the free market.

THE CASE FOR MARKET-BASED INCENTIVES

In any marketplace, the burden of production tends to be allocated to the lowest-cost producers. Thus, the rationale for market-based methods asserts that, if there were a market for pollution control, those who could control their pollution levels at the lowest cost would bear the heaviest burden. This might be encouraged by basing regulation on economic incentive strategies rather than the more convenient command-and-control mechanisms. In U.S. environmental programs, this rationale is behind the latest directions in policy making and regulation.

A central concept behind a variety of such approaches is that all consumers and producers ought to face the full costs and consequences of their decisions. Unfortunately, the free market alone (without any government intervention) does not explicitly compel firms to face all environmental costs. Any polluting activity (by economic definition) will also produce costs that firms do not pay. For instance, structural degradation may result from the action of acid rain caused by fumes from the burning of high-sulfur coal in a distant power plant. Yet the costs involved are not factored into the economics of the owner of the plant emitting acid-rain precursors. The acid rain is called an externality since its costs are external to its producer. Externalities are often cited as the most important cause of environmental problems.

The prevalence of externalities provides incentives for actors in the market to produce excess environmental pollution since, by definition, the cost of the externality is not considered in the price of the product. Accordingly the price of the product (which in the above example is electric power) will inevitably be lower than it should be, since it doesn't take into account the environmental costs. In addition, the low cost encourages overproduction of pollution, since there are no incentives for producers to conserve the resources being damaged by the externality. Thus more negative environmental and economic effects are created than would otherwise occur.

To correct this difficulty, producers must realize the full costs or consequences of each decision at the time it is made. When polluters pay the real economic costs for potential problems or subsequent cleanup, environmental choices become just like other business decisions. If the market mechanisms have been well designed, there will be strong incentives to reduce or eliminate any polluting waste and the pollution will be priced *correctly*. Environmental problems, formerly considered external to operations, will become internalized. Management will begin to factor environmental considerations into overall operating and investment strategies for the business.

Over time, society might achieve higher levels of environmental protection at lower overall costs through the use of a market-based system. An effective economic incentive system would bring the creativity and the full power of the marketplace to bear on finding optimum solutions to potential pollution problems. Market-based systems that utilize economic incentives would allow millions of decentralized decision makers—literally every consumer and producer—to make daily decisions that take into account human use of the environment.

Under an appropriate incentive system, if a producer manages to find a lower-cost way to minimize pollution, the result will be lower costs not just to that company but also to the consumers of the firm's products. Competitors will be compelled to use the latest control technologies or lose precious competitive advantages to others in the industry that are employing more advanced methods. Thus, encouraging the search for lower-cost means for environmental compliance brings benefits for environmental quality and also for consumers.

Furthermore, economic incentives also tend to shift the focus from disposal, dispersal, and remediation toward elimination at the source. Since pollution is a form of waste, finding new processes and operating methods to eliminate it might also add to a company's competitiveness. As federal policies encourage this approach in many industries, it could enhance competitiveness across the entire national economy.

Investment decisions can be undertaken carefully when firms receive proper economic signals from the market. If a new approach saves more money over an appropriate payback period than it costs to install, then a rational manager will find it a worthwhile investment. In a command-and-control regime, the threat of heavy fines, bad publicity, and even jail terms can induce firms to provide for proper transport and treatment of waste. But in a more market-oriented regulatory climate in which incentives are provided to reward waste minimization, management might find it even more advantageous to invest in new processes that eliminate pollution or waste in the first place, since the costs of prevention may prove to be less than those for treatment, disposal, and cleanup. In an ideal system, explicit investors will take into account the useful life of new capital equipment. A firm might invest in a much cleaner processing system, for example, if management believed that the new technology would offer an important future competitive advantage when even stricter environmental standards raised the price for compliance. Such an ideal system would also reward the technology developers, enticing them to produce more efficient, less polluting technology because they would be assured of a ready market for their advances. Encouraging moves in this direction might require incentives that take effect early in the investment cycle, because the expected savings might accrue only over a long period of operations. Examples of early incentives include tax abatements or public/private risk sharing through government-guaranteed loans.

Finally, incentive systems help deliver the maximum environmental protection for the money expended. This is important because there are limited resources available for environmental improvement, and tremendous amounts are already being spent by businesses every year just to meet current requirements. By 1991 the annual cost of compliance with environmental regulations in the United States was estimated to be over $100 billion a year, an increase of 56 percent since 1984. Growing concern over the competitiveness of U.S. industry in global markets makes the concepts behind incentives attractive not just to industry, but to the general public as well, because of increasing worries about jobs and economic conditions.

Typical Market-Based Programs: Advantages and Disadvantages

Most incentive systems fall into one or more of five basic categories: pollution charges, fees, or penalties; tradable permits; deposit-refund systems; removal of government subsidies; and reduction of regulatory market barriers. Compared to other programs, these approaches may actually be less costly for government to administer. Funds could be collected by the EPA, for instance, from auctioning tradable permits to the highest bidder. Marketplace dynamics might also help improve compliance, because companies that do spend the money needed to meet or exceed requirements would be more likely to report any violations by their competitors.

Pollution taxes or fees require each generator to pay a fee per unit of pollutant emitted. While this encourages each generator to cut back on pollution to minimize fees or taxes, it also serves to keep investment costs down, because each pollution generator will also be looking for the most cost-effective methods to reduce emissions.

Fee-based waste-management programs can help consumers to understand and face the consequences of waste-generating behavior. To encourage consumers to modify their behavior, many U.S. cities have experimented with rational pricing of garbage-collection services. In most U.S. communities today, the cost of garbage collection, and many other services, is built into the property tax. This practice hides the source of payment for these services and may distribute costs very unfairly. Using a car as an analogy, it's as if the cost of gasoline were allocated according to the value of each individual's automobile. At the end of the year, each car owner would receive a bill for gasoline based only on the value of the vehicle, without regard to how much fuel the owner had actually used. Under such a system, there would be no incentive for an individual car owner to conserve fuel or pay for costly tune-ups. The biggest gasoline users would gain the maximum benefit from this odd sort of gas tax. Yet allocating garbage collection costs according to property evaluations can be viewed in a similar light. It fails to encourage each property owner to minimize trash generation. The need for some clear incentive to cut back on garbage becomes especially acute if the community hopes to encourage the public to recycle.

Fee-based solid waste collection services use a per-bag or per-can charge to help the customer connect the costs associated with waste management directly to the act of setting out the garbage. Surveys have indicated that the programs have not only increased recycling and composting, but have also influenced purchasing behavior, to avoid products that use excessive packaging, for example. Surprisingly, the studies have shown that most residents actually favor this strategy. Practical problems face regulators when setting up such a system, however, since unscrupulous dumpers might throw their garbage onto the piles of others or dump it illegally. Also, the costs buried in property taxes are not likely to be refunded once such a system is initiated. The waste charges would only add new costs to the consumer, a practice that makes selling the concept politically difficult and may keep programs like this from implementation.

Other programs instituted by some states levy charges on hazardous waste disposal. Implementation and exact charges vary widely, with most fees collected at the disposal site. To illustrate differing practices, in 1984, Wisconsin charged $14 per ton while its northern neighbor, Minnesota, demanded more than five times that amount. Even though these fees tend to influence businesses to cut their output of waste, they also raise costs because not all trash can be eliminated. In fact, the major motivation for the imposition of such fees, reports a senior staff economist from the Council of Economic Advisors, is not so much to reduce waste as it is to find a new source of revenue, even though the money collected may be earmarked for activities that promote environmental quality.

Even though fees do tend to force businesses to make better production decisions, they still fail to solve several underlying problems. It is very difficult to estimate the right fee or tax to charge to obtain the most cost-efficient environmental solutions. While some economists claim that they can estimate relevant pricing elasticities and thus establish tax rates based on marginal costs, marginal benefits, and resource reserves, others assert that there are so many guesses and unknowns involved that outcomes could easily be far from expectations. There is really no way to accurately predict, a priori, how much pollution abatement will be achieved for any given schedule of fees or other charges. In addition, such schemes entice the wrath of environmentalists since they do not place any real cap on the aggregate level of pollution.

An intriguing alternative that is, in a sense, a mirror image of a pollution charge, is a system of tradable permits. By substituting permits that restrict the amounts of allowable pollution for fees on whatever is generated, this approach does provide the desired cap on the total amount of pollution. Since permits are tradable, polluters that can reduce or eliminate waste beyond the amount they are assigned can sell their excess for a profit to others that may find it too costly to meet their assigned targets. Therefore, incentives exist for polluters to develop technology that cuts emissions.

Allowing this trading in permits achieves two important objectives. First, regulators place a cap on total emissions since a finite aggregate amount of pollution is allocated on the permits issued. In addition, as soon as trading takes place, total costs over the population must be decreased because lowering costs is the force driving the market. Another important benefit is that the EPA will be able to lower the overall level of pollution over time by removing permitted amounts from the market as technology improves.

This exact scheme was used in what is one of the greatest success stories in the history of U.S. environmental regulations. The United States successfully phased most of the lead out of gasoline, reducing it to 10 percent of its original level in a period of only 3 1/2 to 4 years using a system of tradable permits among refineries. By means of this approach, the EPA, which supervised the permit system, estimated that the goal was accomplished at a cost saving of some $200 million a year over regulating by means of strict performance standards without allowing trading.

Now, with new amendments to the Clean Air Act, the most comprehensive approach that has ever been tried with tradable permits is just beginning. Initiated by the Bush administration and approved by Congress in 1990, this legislation sets up a tradable-permit program for sulfur dioxide and, later, nitric and nitrous oxides, which are considered to be precursors to acid rain. Again it has been estimated (the results depend on how the estimates are calculated) that allowing tradable permits will achieve the desired control at only half the cost—a total savings of about $3 billion a year over what otherwise would have been needed.

The air permit plan allows facilities to earn emission credits that can be applied to future emissions or to other facilities if they reduce emissions below the federal standard or ahead of the timetable set by law. Currently, power plants are the major targets of this legislation, and permits are indexed to energy output. By 1995, plants emitting more than 2.5 pounds of sulfur dioxide per million British thermal units (BTUs) of output will be required to cut back to the 2.5-pound level, while those that exceed the standard will be allowed to sell excesses to plants that don't meet the limits. On January 1, 2000, the standard will be decreased to 1.2 pounds per million BTU. Then allowances for those exceeding the standard will be given only to those plants exceeding a 0.8-pound-per-million BTU threshold.

While tradable permits may be suitable for creating market-based incentives for some types of pollution, we need different mechanisms to deal with other forms of waste. Deposit-refund systems provide another method for introducing incentives for environmental management in cases where reuse or recycling is feasible. Several states in the United States, along with some Canadian provinces and European countries, have instituted deposit-refund systems for containers such as bottles and cans. Although this application may not be the best example for the deposit-refund idea, the concept might prove quite cost-effective for certain types of containerized hazardous wastes such as lead-acid batteries.

Car batteries dumped into landfills are believed to be a major source of the poisonous lead that leaches into groundwater. Batteries that find their way into incinerators add to the lead burden in the air. While each of these methods of disposal is illegal now, batteries often reach these facilities undetected in household trash. Even though 70–90 percent of battery lead is already recycled in the United States each year, depending on fluctuations in the price of virgin lead, essentially 100 percent is achievable at what might be a modest cost. A refund scheme, structured around the existing infrastructure, would promote a self-monitoring program that would induce almost all consumers to return old batteries for refunds rather than dumping them—a practice difficult to either detect or deter.

Regulators must be careful how these types of incentives are used. Unfortunately, the deposit-refund system as it is applied to beverage containers is not as useful as it may seem. Bottle and can deposit laws remove the most valuable materials from the curbside recycling stream, cutting into the profits of recycling programs. An important contribution to the profitability of recyclers comes from sorting used containers made of materials such as various metals and glass from the waste stream. Besides hindering recycling efforts, using deposit-refund systems can cost significantly more than curbside recycling. In fact, the California Department of Conservation reports that its programs recycle a total of 9.2 billion containers at a cost 94 percent lower than that of deposit-refund systems in 9 other states.

Another approach to engendering economic incentives in some other environmental areas is to remove existing regulatory barriers that prevent open-market activity. One example of this is allowing those who have water rights in California to sell them to others. Some farmers might be able to obtain funds to invest in more water-efficient distribution systems by selling a portion of their water rights to urban areas. This might lead to more efficient distribution of water without requiring costly and disruptive new water development projects.

In 1988, the Bush administration formed a task force to look more deeply into market-oriented policy options, and particularly to examine possibilities for harnessing market forces to better protect the environment. The result was Project 88, a public policy study sponsored by Senators Timothy E. Wirth of Colorado and John Heinz of Pennsylvania. A group of more than 50 environmental specialists and policy makers, led by Dr. Robert Stavins, an economist and professor of public policy at the John F. Kennedy School of Government at Harvard University, performed a comprehensive analysis of 13 environmental and natural resource problems facing the nation, and recommended 36 policies for stimulating market forces to help deal with the problems.

The Bush administration embraced the market-incentives approach, and William K. Reilly, then administrator of the EPA, established an Economic Incentives Task Force to investigate the potential for market-oriented policies in all areas of the agency's jurisdiction. The tradable-permit system for dealing with acid rain, described above, was recommended by the Project 88 Report. The task force was expanded to over 100 environmental specialists to do a follow-up study, again under Dr. Stavins, which resulted in a Project 88-Round II Report in 1991. This second report focused on the specific issues of global warming, waste management, and public land management.

Consequently, with numerous policy recommendations backed up by in-depth studies, it is likely that many areas of environmental regulation will turn more to market incentives in the future. The transition away from command-and-control approaches is likely to be gradual, however, because of institutional obstacles. Regulators, for example, feel uncomfortable when they cannot depend on concrete, well-defined measures for enforcement. Environmentalists tend to oppose regulatory approaches that do not include explicit target levels for improvement. Sometimes industry may be uncooperative in responding to inducements. Some companies with low pollution levels, for example, may choose not to sell tradable rights to other firms that need them. A period of experimentation may be needed to help policy makers learn how to shape market incentives to the realities of the marketplace and to achieve the desired environmental progress.

DIFFICULTIES OF IMPLEMENTING MARKET MECHANISMS

There are many cases where existing marketplace incentives do not encourage environmental stewardship in practice. For example, newsprint is produced from virgin wood pulp at costs roughly comparable to those for reclaiming and de-inking discarded newspapers, but this is only because many of the true costs to society are not included in the prices actually charged for newsprint. Discarded newspapers constitute a significant portion of the debris filling up scarce landfill space, for example, and logging can degrade the environment through soil erosion, damage to fisheries and water supplies, and destruction of wildlife habitat. Yet these social costs are not currently charged to lumber producers and then, subsequently, to publishers and newspaper buyers. The costs of disposing of discarded packaging similarly have not been passed back to producers or consumers. It remains to be seen how widely market incentives will be developed to reflect such costs back to the waste generators.

Unfortunately, even though economic theory may suggest that incentives can encourage a quest for optimal solutions, evidence shows that, in many cases, U.S. firms and consumers are not as sensitive to price as might be expected, and therefore they may be less inclined to take on environmental projects. A very disturbing example of a real-world problem associated with market mechanisms is provided by the energy industry.

U.S. industry is about half as energy-efficient as West Germany and Japan, for example. This might be explained partly by lower energy prices in the United States, but a substantial portion of the difference seems due instead to firms' indifference, probably due to the traditional view of waste management as an externality. One study revealed that utility customers in New York State, including commercial and industrial facilities, could save 40 percent on their energy bills through conservation alone.

Public utility commissions in some states are sensitive to this indifference and have offered incentives to utilities if they can help their customers conserve energy. In those states where demand is close to reaching generation capacity, utilities also gain when their customers save energy. All of the savings free up capacity needed to help the utility avoid building new generating facilities, which would have to go through a costly permitting process and be built to expensive new standards. In cases where the capacity shortage is critical, utilities have even offered to pay for everything from audits and design to equipment and installation. However, most utility-sponsored programs generally involve helping clients to make investments in energy-efficient devices or equipment that would offer energy savings. In theory, market forces should induce such investments without encouragement from utilities, but, as in New York State, many potential savings remain unrealized.

Much of this indifference could be due to a lack of adequate information and understanding by decision makers. Stronger programs for education and training might reverse this trend, and some environmental legislation does include funds for just this purpose. During Jerry Brown's tenure as governor of California, a group of energy specialists with industrial background was set up by the state, and they worked with both the power companies and industry to encourage conservation. Utilities were allowed to offer off-peak pricing for electricity, for example, and then heavy users of electric power, such as cement plants, were encouraged to do much of their processing at night to cut costs. When visiting large industrial plants, this group found that very simple steps, such as cutting back on heating and air conditioning over weekends and holidays, could often produce big savings.

Government subsidies often provide another barrier to proper market functioning. In some cases they actually encourage environmental damage. One example is the subsidization of timber sales in the Northwest, where the U.S. Forest Service does not even earn enough to pay for the cost of making timber available for harvest by private firms. It is estimated that current timber subsidies cost the federal treasury about $400 million a year, while timber harvests cause many types of environmental degradation. Reducing the subsidies could discourage excess cutting and reduce the resulting environmental damage to watersheds and wildlife habitat while increasing net federal revenue. It would also raise lumber prices, thus creating markets for alternatives. New types of products, such as microlaminated beams that can be made even stronger than traditional timbers and extruded framing structures made from cellulose fibers, would be more widely used. A wide variety of plants could be used to produce fibrous materials that could take the place of raw lumber. Similar subsidies for flood control, grazing on federal lands, mining rights, and agricultural price supports also create disincentives for environmental protection.

CITIZEN EMPOWERMENT AND THE RIGHT TO KNOW

A final strategy that policy makers have used to promote environmental policies is the use of public dissemination of environmental information. This action forms a key component of the environmental management process. When the public is informed about an issue, and even more so when citizens participate in the decision making, the outcome is less likely to ignore essential considerations regarding the well being of human health and the environment. These mechanisms may seem inconsequential at first blush because they typically have few binding requirements, but they do tend to invite political review, which is often much more effective than other mechanisms.

The National Environmental Policy Act of 1969 was the first environmental act to adopt this tactic by requiring the now infamous Environmental Impact Statement (EIS). The EIS applies to all "significant" projects that are funded by the government. It requires the project's proponent to analyze the effects of the project carefully and contrast them with other viable alternatives—including the "no-build" alternative—and provide the results for public inspection. The law does not provide any direct legal recourse for those who would reject the project on environmental grounds; it only requires that the proponent do a thorough and honest job of exploring the alternatives.

Although the EIS is a simple requirement, it has resulted in a shift in the types of projects undertaken by the U.S. government. Many public works construction projects, such as dams and other water related works, have been shelved as the result of the imposition of an EIS and the resulting furor expressed by the public. One reason this approach may have worked to change the nature of public construction so successfully is that a comprehensive impact statement can provide good arguments for political figures who can exploit the process for the sake of the publicity it generates. In any case, enough public clamor usually encourages the proponents to either abandon or change the project, either through legal action or resulting political pressure.

Another technique, commonly called "Right To Know" (RTK), would provide information that effectively shifts the balance of power toward citizens by allowing them to find out about the risks that they are exposed to due to operations of the firms in their vicinity. A typical RTK law would require a firm to notify the closest towns of all hazardous and toxic chemicals it currently uses and provide a plan for the protection of the local population in the event of an accident. This requirement has promoted better environmental practices among companies since they must audit their own practices (increased self-awareness) and distribute information (which may become evidence in potential lawsuits). In many instances, the collection and dissemination of this information, such as in siting new nuclear power plants, has sensitized the companies to the threats from environmental lawsuits to the extent that they encourage practices such as waste minimization.

DIVERSITY IN PERSPECTIVE ON ENVIRONMENTAL PROTECTION

While many different policies and types of programs are used in the United States for environmental protection, none is without flaws. Each has drawbacks that cause the system to break down in one way or another. Therefore, no single method is inherently better than another. Furthermore, environmental policies are never judged simply by the environmental criteria that are primarily discussed here. Many other considerations, especially political pressures, influence the process of policy making and regulation, and these must not be ignored.

These diversionary forces within the system may lead to policies that may at first seem illogical, but may actually be inevitable. Environmental statutes have been the victim of the redistributive nature of the political system with narrow interests of different sections of the country resulting in patchwork solutions that are not the most beneficial for the nation's environment. For instance, an attempt was made in Congress to *forbid* new electric power plants from switching to low-sulfur coal in order to reduce pollution discharges. Although such a switch would benefit the environment and result in tremendous cost savings, the antiswitching policies were aimed at trying to protect the jobs of a relatively small number of high-sulfur coal miners. Similarly, the federal subsidies for certain sewage treatment plants remain even though the plants may have only marginal effects on water quality. Unfortunately, some of these plants amount to pork-barrel projects that politicians simply can't resist. Another trouble spot lies in non-point-source pollution, such as the runoff from farms and orchards, which is perhaps the most destructive water-quality problem in the United States today. Yet the problem has not been seriously addressed primarily because powerful interests have been able to resist new regulations that would restrict their operations.

The widely divergent views held by the public also serve to complicate the resolution of environmental problems. Continuing debate, which has become more heated as the green movement gains momentum, centers on methods used by the government and on what constitutes satisfactory levels of environmental contamination. These clashes often involve groups with radically different views on these issues.

At one end of the spectrum are ardent environmentalists. Many members of this group feel that excess pollution and overdevelopment are steadily eroding the quality of human life in settled portions of the globe. Furthermore, they see undeveloped areas rapidly shrinking while those regions of pristine wilderness that do remain are assaulted by the steadily increasing air- and water-borne effluent of encroaching civilization. Environmental purists want to see the remaining wilderness preserved and the abuses associated with past development rectified by returning any effected air, lands, and waters to as close as possible to their "natural" condition. At the opposite extreme are avid developers who feel that the needs of economic expansion and increased employment far outweigh the "trivial" concerns of "tree huggers." They see ever-expanding government regulation, with consequent delays and high costs for compliance, as an economic drag that diverts capital from productive investments, eats into profits, and eliminates jobs.

While there are few who actually admit to either of these absolutist perspectives, there are powerful advocates and groups with positions staked out close to each extreme. Governments have tried to play the arbiter through regulation, with results that typically satisfy neither of the extreme groups.

The rule-making and enforcement process sometimes fails as well. Regulators may have to fight powerful special interests that can afford lawyers skilled at using the courts to delay compliance while using political influence to induce the environmental agency to moderate demands on their clients. Furthermore, some of the tasks assigned to the EPA by Congress, which superficially seemed to be simple and straightforward, have turned out to be much more difficult than expected. For instance, coming up with definitions that clearly specify which substances are truly hazardous resulted in countless bioassays, test specifications, public hearings, and industry lawsuits. Yet the controversy continues. Such determinations turned out to require value judgments because of scientific uncertainties and complexities that had not been foreseen by either Congress or the EPA.

Even with all its faults, the U.S. environmental management system is perhaps the most advanced in the world. Many nations look toward the United States for leadership in environmental policy and regulation questions of local, national, and even international significance.

MAJOR AREAS OF U.S. POLICY AND REGULATION

Environmental policy, laws, and regulation are provided by the legislative, executive, and judicial branches of the federal, state, and local governments. Although many states and localities have provided for their own measures of environmental protection and do have a substantial impact on the implementation of much of the nation's environmental policies, the federal government is most responsible for the development and prosecution of major policy moves.

There are three major clusters of U.S. federal environmental policy and regulation that apply to the environment: (1) airborne waste, (2) waterborne waste, and (3) solid and hazardous waste. Although the government tends to make much finer distinctions, this level of aggregation is sufficient to grasp the concepts that policy makers use to guide current decision making.

Airborne Waste

Before 1970, airborne waste was largely unregulated, unless it constituted a direct and objectionable public nuisance. Relatively small-scale air pollution control efforts that were initiated in many cities resulted in air pollution restrictions designed to control smoke and soot (particulates) problems from furnaces and factories. These laws were limited in geographic scope and usually only prescribed simple measures such as limiting operating hours.

Early federal legislation aimed at reducing air pollution was both vague and impotent. Throughout the late 1950s and during the 1960s, legislators provided appropriations for research, technical assistance, and training for state air pollution control personnel, but little more. For instance, the original Clean Air Act of 1963 provided for permanent federal support for air pollution research and channeled funds into the development of state air pollution agencies, but it didn't create any concrete requirements for air quality or discharge control.

The federal role in airborne waste control expanded widely in 1970 after the passage of a new set of amendments to the Clean Air Act. These gave direction and power to the newly created EPA by specifying the goals of the new air pollution control program and describing the means of attaining these goals.

The most significant goal of the new program was the development of the National Ambient Air Quality Standards (NAAQS) that were intended by Congress to protect the public from the adverse effects of air pollutants in the ambient air. Although the strategy was a milestone, it was seriously flawed in at least two respects. Congress implicitly assumed that there were "safe" levels of air pollutant concentrations, below which there was no physical response. The growing body of physiological and toxicological evidence seems to suggest that there actually may be *no safe level of exposure for many pollutants*. Therefore, the regulators were (and still are) faced with a dilemma: absolutely safe implies shutting down all energy production, commerce, and industrial activity, while allowing pollutant emissions places some (hopefully small) portion of the population at risk.

The second flaw in the program was that of practical enforcement. The NAAQS were set without regard to the costs of implementation. Therefore, there was no provision to trade off the costs of attainment vis-à-vis budgetary requirements. Although necessary from a practical implementation standpoint, this stipulation was not to be a consideration during standard promulgation.

Nonetheless, the EPA was instructed to provide performance standards for mobile and stationary sources. In the case of mobile sources (cars, trucks, buses, motorcycles), Congress itself actually determined the emission standards by calculating emissions reductions needed to meet ambient air quality standards at the time. Congress also directed the EPA to establish emission standards for stationary sources (power plants, factories, refineries, and other industrial facilities) using a complex system of technology-based criteria and considering the ability of industries to afford the treatments.

By 1990, many involved parties had recognized that the problem of overall program costs was serious enough to warrant several changes in the act. The annual compliance cost had risen to approximately $50 billion, and a substantial portion of this investment was made inefficiently. Those arguing for change made the point that the economics of every source is different, and controlling equally at sources with both high and low costs of control was driving the cost of pollution control up unnecessarily. Most agreed that the total cost of pollution control could fall while aggregate pollution control remained constant or even fell. This provided the impetus for the 1990 amendments to the Clean Air Act.

The most important change in airborne waste control policies due to the new amendments was the development of the tradable-emission credit system, which allows real markets for pollution-control expenditures to develop. Facilities that reduce emissions below the standard or ahead of the timetable set by law will be allowed to sell emission credits to another facility. Sales of credits should level the cost of compliance with environmental requirements and provide incentives for facilities to undertake pollution-control projects. This market-based approach to the solution of environmental problems is likely to be the direction of many different laws in the future. Air quality is revisited in Chapter 4 in a discussion of technological implications of regulatory actions.

Waterborne Waste

Waterborne waste regulation, like airborne waste regulation, was largely ineffective until the late 1960s or early 1970s. Several different federal laws that were intended to promote water pollution control and establish water quality standards were either ineffective due to enforcement mechanisms or funding provisions. Even though the Water Pollution Control Act of 1956 did authorize federal grants which could pay up to 55 percent of the cost of construction of wastewater facilities, the legislation left decision makers with wide discretionary powers that ultimately were responsible for few meaningful contributions to water quality change.

The Water Quality Act of 1965, which for the first time required states to establish ambient water quality standards, was ineffective due to the inability of enforcement personnel to penalize sources of pollution. Ambient standards provided no mechanism for regulators to determine who was polluting and how much, if any, the ambient standard was exceeded. Furthermore, enforcement actions were mostly the responsibility of the states, which were accommodating at best and often bowed to industry pressure to maintain good relations with native firms.

Congress responded to these concerns with the Federal Water Pollution Control Act of 1972 and its extension, the Clean Water Act (CWA) of 1977. These acts were marked by a new direction in the federal water pollution control program: Congress set a subjective but measurable goal of providing "fishable and swimmable waters" in the United States. The acts established technology-based effluent standards, determined by the EPA for different classes of polluters. This eliminated the need for regulators to determine ambient standards or calculate how much of a certain pollutant a particular body of water could assimilate in a given period of time. They also recalled the authority for oversight from the states and shifted this activity to the federal government. The legislation was also intended to encourage projects, so the act raised the maximum federal contribution to plant construction costs to 75 percent. Finally, the program established a system of discharge permits that were intended to provide regulators with a mechanism that allowed the use of higher standards in areas that were more sensitive to pollutants or heavily polluted. Permits, required for every discharger in the country, became the primary means used in the United States to regulate the discharge of waterborne toxics.

A final significant class of water pollution called "non-point-source" pollution is quickly becoming an important focus of regulatory action. It is called non-point-source pollution since it cannot be traced to a particular industrial or commercial polluter, but comes from sources such as storm runoff or agricultural, industrial, silvicultural, construction, and mining drainage. Many consider the control of non-point-source water pollution as the most significant water quality problem in the United States since it accounts for more than 50 percent of all pollution currently entering U.S. water bodies. Non-point sources are also regulated under the Clean Water Act, but the program has been hindered by inattention and neglect since it is overshadowed by the more politically motivated, concentrated, and visible water-treatment-plant construction program.

Solid and Hazardous Waste

The first modern-day federal response to problems associated with solid waste management was the passage of the Solid Waste Disposal Act (SWDA) of 1965, which authorized modest technical assistance to municipalities and promoted research into areas such as recycling, waste processing, resource recovery, and residue disposal. It also provided for the promulgation of guidelines for solid waste collection, transport, separation, recovery, and disposal systems.

In 1976, Congress amended the Solid Waste Disposal Act with an influential set of amendments called the Resource Conservation and Recovery Act (RCRA) of 1976. The act was a needed but hasty response to the problems of both solid and hazardous wastes. At the time, neither Congress nor the EPA realized the scope of the program that would inevitably result from this act, especially regarding hazardous wastes.

The RCRA made notable progress in the area of solid waste: it gave broad federal regulatory power to the EPA; it emphasized the use of resource recovery, recycling, and waste-to-energy techniques; it mandated procurement policies that would utilize more recycled and recovered materials; it attempted to shift the burden of costs of landfilling to users. It also established substantial research, development, and information requirements on a variety of subjects, such as the magnitude of waste produced; the social, economic, and environmental consequences of different measures; and the appropriateness or fairness of taxes, fees, and other regulatory mechanisms. What it did not do was to mandate federal standards or provide a national policy for municipal waste.

The 1980s also saw a marked reduction in the amount of federal attention paid to solid wastes, although the solid waste troubles of the nation continued to be newsworthy and important to the public. This trend is especially obvious when one examines the noticeable lack of funding given to federal-municipal solid waste programs in the 1980s. This persisted even when stories appeared in major news media about trains, barges, and trucks carrying refuse and other unsavory cargoes from place to place, with nowhere to unload. Many landfills shut down, some of them recognized as sites that could endanger the public. By 1986, the Office of Technology Assessment (OTA) reported that 22 percent of all Superfund sites were municipal landfills.

Nonetheless, in spite of federal inattention, solid waste management policy has become more refined during the last decade. The EPA is now attempting to take a more holistic approach to solid waste problems by applying what it has called "integrated waste-management" principles. In practice, the EPA views the policy largely as an interdependent stream of waste-management activities in a comprehensive program. First, municipalities should reduce the volume and toxicity of waste as much as possible and recycle or compost any valuable or organic waste. Then, they should remove any noncombustible portions to leave waste that will provide good fuel for energy production. Where possible, materials that produce toxic by-products when burned or that leave toxic constituents in combustion ash (in amounts that could be considered hazardous) and materials that interfere with the combustion process (for example, because of their physical composition) should also be removed. The combustible fraction should be burned, and only previously separated materials and remaining combustion ash should finally be landfilled.

While solid waste programs essentially languished, hazardous waste legislation became an important focus of the EPA. The EPA subjected generators and transporters of hazardous wastes to a system of record keeping that was intended to firmly fix generators with specific responsibilities regarding the final disposal of the waste they generated. Congress also biased the program against the land disposal (burial or landfilling) of hazardous wastes.

During the implementation of RCRA, Congress also began to turn its attention to the problems associated with the numerous sites where hazardous waste had been disposed of in the past. This process was sparked by the national tragedy associated with the now-famous residential area called Love Canal where dangerous chemicals were found to be leaking into the basements of residents and contaminating the drinking water of the community, which had knowingly been built on the site of an old hazardous waste dump.

Another section of RCRA calls for strict regulation of underground storage tanks containing hazardous substances, and an increasing amount of the EPA's attention may be turned to this area. The agency estimates that there are nearly 2 million underground petroleum tanks at 70,000 facilities and 50,000 tanks containing other substances at 30,000 facilities that fall under the Hazardous Solid Waste Amendment (HSWA) provisions. After analyzing these data and making an assumption that the life expectancy of such tanks is about 15 years, the EPA estimates that up to a fifth of these tanks may already be leaking.

Since RCRA was inadequate in dealing with the problems associated with past practices, Congress passed the Comprehensive Environmental Response, Compensation and Liability Act of 1980 (CERCLA) or Superfund Act which was intended to clean up old hazardous waste disposal sites around the United States. Under the program, thousands of existing hazardous waste sites have been identified, evaluated, and prioritized for cleanup. The government also determines who is responsible for the contamination at the sites, and the regulators inform those responsible of their findings and then request that these potentially responsible parties (PRPs) implement a cleanup. If the PRPs fail to clean up the site, the government administers the cleanup and sues the PRPs for all costs incurred.

The Superfund program differs from most other programs in that penalties are paid after the environmental damage is already done. This doesn't allow polluters to use foresight to weigh the costs of their polluting activities. In legal terms, this is an unusual precedent: firms are liable after the fact even though they may have been in complete compliance with existing laws and current best practice when they caused the environmental pollution. Although firms cannot change poor past practices now, the high costs and negative publicity involved in dealing with Superfund sites does serve as a strong inducement for corporations to alter their inadequate waste-management practices, eliminating hazardous wastes when feasible and properly disposing of any that can't be avoided.

The term Superfund is derived from the Hazardous Waste Trust Fund (a superfund) to be built up through taxes on crude oil and over forty commercial chemicals. These funds would be used for making emergency response to spills, for cleaning up sites where a responsible party could not be identified, or for cleaning up sites where those who had allegedly caused the contamination refused to pay for remediation. Funds are also available to take any such recalcitrant dumpers to court and force them to pay for the cleanup.

Some observers point out that about half of the Superfund money so far expended has been used to pay lawyers and lobbyists rather than to pay actual cleanup costs. Unfortunate as this fact may seem, the threat of being called to account for hazardous waste does have one intended effect on the views of executives in many industries. Management has increasingly begun to consider the costs of hazardous waste and their ultimate disposal as an important concern for their business. Since executives recognize that their organizations can be held liable for any damages resulting from improper disposal, these ultimate costs are being internalized into everyday decision making by the firm. The result in many industries is a much more careful look at not just disposal of waste, but entire processes that create the waste in the first place.

However well the program has encouraged better management of current waste, it has also been roundly criticized for its lack of progress in reaching its intended goals of remedying past problems. In current Superfund cleanup efforts, for example, the EPA must approve the technology chosen to perform remediation and then must certify that the task has been satisfactorily completed. This approach can impede the emergence of promising technologies, eventually discouraging those responsible for a cleanup from considering new technology due to the difficulties associated with obtaining necessary approvals from the EPA.

LEARNING FROM FOREIGN EXPERIENCES

Knowledge gained form the experiences of other nations can also help U.S. environmental efforts. Many nations use different approaches to achieve sustainable development while curbing environmental degradation. We can derive new concepts for improving environmental policies and regulation in the United States from successful programs in other countries. These types of programs may be harbingers of new policies and programs in this country.

The experiences of the European Community (EC) in attempting to develop and implement environmental policies are particularly instructive. Policy development went through three distinct phases, starting with efforts to break down trade barriers and moving steadily toward greater economic and political integration. The first phase did not really involve environmental policy, but dealt rather with some rules that served as a form of trade barrier between the individual European states. This approach can be termed "negative integration." Negative integration focuses on abolishing discriminatory national rules and legislation. By contrast, "positive integration" involves adopting common and coordinated policies to promote economic welfare in a concerted fashion. Examples of some of the negative integration issues dealt with over the period from 1964 to 1975 include establishing a uniform system for the labeling, packaging, and classification of hazardous waste and regulation of the noise and exhaust levels of vehicles.

The second phase of EC environmental policy making began in 1971 with steps toward a communitywide commitment to environmental protection. To institute this policy shift, officials had to reinterpret some of the articles in the original treaty setting up the Economic Community, allowing them to deal with communitywide environmental issues under harmonization provisions that had been originally intended to cover only economic matters. An environmental commission was set up, and it developed an initiative that was endorsed by the member governments in 1973. The program laid out by the commission included action plans to cover all types of natural resources threatened by industrial development and economic expansion. This initiative started a move away from negative integration issues and laid the groundwork for a later shift to a positive integration approach. The resulting legislation did not look toward the future, but dealt instead with existing problems, such as disposal of waste oils, transborder waste shipments, dumping waste at sea, and the disposal of polychlorinated biphenols (PCBs) and PCTs. Individual states passed their own laws within the overall framework set up for the total community, and legislation dealt almost exclusively with existing products while new products were almost totally ignored. No environmental sector was completely covered by EC policies, and perhaps only a fifth of the relevant problem areas were covered at all. Priorities were not defined, and only a patchwork of areas were covered by local regulations within the various states. There was also no enforcement power at the community level.

While efforts were made toward positive integration over the next few years, there was little progress because any decisions had to come out of a unanimous consensus of all members. So, those nations, particularly in southern Europe, that lagged in environmental progress could stall any initiatives that would impinge on their national sovereignty. In 1986, as part of the Single European Act (SEA) that set the end of 1992 as the deadline for the elimination of all customs and trade barriers across member nations, the unanimous-consensus requirement was changed to majority rule. This led to the third phase of environmental policy making, during which there has been more progress toward integration and harmonization than in the previous thirty years. Since the greener states in the community could now push through environmental policies more easily, a variety of new concepts were brought forward in the late 1980s, including waste minimization, life-cycle analysis, clean technologies, and environmental auditing. More active policies now being considered are much more comprehensive and are aimed at getting industries to internalize the total cost of the pollution and waste they generate.

Policy has been propagated throughout the EC by means of four Environmental Action Plans (EAPs) initiated over the past two decades. The first one, begun in 1972 and covering the 1973–75 period, set out principles that determined the course of EC policy over the next decade and a half. Among the main political principles stated in this first agreement were:

1. Prevention is more effective than ex-post-facto treatment.
2. Polluters are financially responsible for the pollution they create.
3. The most appropriate geographic decision-making level will be sought for each type of action.
4. Pollution across boundaries must be prevented.
5. Assessment of environmental impact of official action is needed.
6. The states are allowed to enact more stringent national regulations.

These principles proved not to be very effective due to their vagueness and some conflict among them. The most blatant conflict came from the ability of individual nations to set up regulations stricter than community standard and thus to create trade barriers. Along with the principles, an action program was also passed with the stated objective of setting out quality standards for airborne and aqueous pollutants. Well-defined actions were proposed to control water pollution from various industries, including pulp and paper and the titanium dioxide industry.

The second EAP extended from 1976 to 1981, and, although the first program's policies were repeated, the new initiative was marked by a shift from pollution control toward pollution prevention. Issues such as waste management, nuclear energy, and noise control were addressed, but to a limited extent. This EAP also set up environmental impact assessment, the ecological mapping of the EC and a move toward a resource economy, concentrating on recycling as its first goal.

A third EAP, from 1982 to 1986, served to consolidate EC environmental policy in conjunction with policy developments in the member states themselves, while reiterating the need to implement five major directives concerning the aquatic environment, titanium dioxide, sulfur dioxide limit values, toxic waste, and dangerous substances. The role of a Council of Ministers was strengthened during this period, by approving the general direction of the environmental program and developing a priority list—by setting priorities, the Council effectively gave the policies considerably more binding power.

The fourth EAP differed from the previous three by giving new legislative powers to the EC to effect proposed recommendations. It was part of the Single European Act—fixing objectives for future policy for protecting the environment—and recognized environmental policy as a component of other types of policy. The result will be a single European-wide environmental policy based on the prevention of pollution and a multisectoral approach to prevent the transfer of pollution from one medium to another. There are three basic themes in the fourth EAP:

1. reduction of waste
2. recycling and reuse
3. safe disposal of unavoidable waste

Since its beginning in 1973, EC environmental policy has progressed further than its critics, or even supporters, initially believed possible. Even though most directives and regulations that have been adopted tend to be narrow in scope, tackling small issues, there has been some legislation with substantial impact, such as limits on automotive exhausts and sulfur dioxide emissions. There is a divergence between the broad, long-term objectives of EC rule makers and the local interests of regulatory bodies within each member state. Many environmental officials remote from central decision making are not even aware of EC directives in their day-to-day work. EC directives are generally vague, which is necessary for communitywide adoption, but the resulting application may be so inconsistent as to defeat the overall purpose of the rules. Discrepancies in the degree of compliance by member states also can undermine the debate process. France, for example, may object to certain strict regulations because of fears of the impact on its industries and government spending, while Italy may not object because its officials know they would be unable to apply the principles anyway because of a lack of strong central authority over semiautonomous regions.

In spite of such flaws, an effective environmental framework has been established, and what is now needed is an agency with the power to enforce EC directives, analogous to the EPA in the United States. The fourth EAP will probably be followed by a fifth program which might incorporate such an institution. It is likely to be a new European Environmental Agency—an agreed-upon but still nonexistent entity. While such an agency has not yet been created, there are many ideas and plans that have been proposed to coordinate European waste-management activities, one of them being a unified ecological labeling system being urged by many companies and countries. Creating an agency that can effectively carry out such plans will be the next challenge for the EC.

INTERNATIONAL ACCORDS

International accords will increasingly impact the conduct of all nations and are likely to become a significant source of pressure for environmental change in the future. Under today's international law, there is no real sanctioning power available for one nation's government to regulate either global or transborder intrusions from its neighbors since nations are considered sovereign. For years, many different nations have exercised power over others by providing incentives and disincentives in the form of conditional loans, threats, and other international tactics. Although such individual negotiating will undoubtedly continue, the need for greater international cooperation is becoming evident as more global problems—such as acid rain, ozone depletion, and global warming—are recognized as potentially significant threats to the health of the future environment, not just regionally but indeed for the whole planet.

The crafting and implementation of world-scale environmental policies and management schemes is much more difficult than domestic ones. Many other issues are tied to environmental problems, such as the growth of populations and economic development goals of individual countries and regions. Needless to say, many developing countries resent the implication from outsiders that they must change their management methods due to outside concerns. Developed countries also resent the implications that they should pay for the pollution programs they suggest. The developing countries also point to resistance of governments in industrialized nations, including the United States, against modifying their systems to reduce pollution, save energy, and so on.

No area of environmental concern better illustrates the complexity of reaching international accords than the global warming issue. The potential for planetary environmental disaster posed by the buildup of greenhouse gases has become a cause for great concern all over the world. But, as is so often the case in the environmental arena, the issues involved are not at all clear-cut. Theory indicates that, as emissions from the burning of fossil fuels collect in the troposphere, the atmosphere will begin to act like a greenhouse, trapping solar heat and warming the earth. There is a huge scientific (and policy) dispute, however, over the application of the theory to atmospheric models so as to forecast the extent of the impact of a buildup of greenhouse gases. Some believe that there will be only minor effects, with average temperatures rising only 1.5 to 2 degrees Centigrade (°C) in a century. Such a minor change would have insignificant effects. Others fear much more dire consequences. They predict that temperatures might rise by perhaps 4–5 (°C) by the end of the next century with potentially devastating results. Evidence for the buildup of these gases comes mostly from ice core samples taken from deep within polar glaciers and is supplemented by some atmospheric measurements. One remarkable Antarctic sample provides concentration data going back some 160,000 years. It shows that in the 120,000 years before the industrial revolution began (about 1865), carbon dioxide levels scarcely changed at all, remaining at about 280 ppm in the atmosphere. Since that time there has been an increase of roughly 25 percent, to about 355 ppm. The rise in concentration of greenhouse gases over the period from 1865 to 1960, some 100 years, was about the same as the increase in the past 30 years, and the same rate of increase is true for methane. The concentration of methane, another greenhouse gas, is now about two and one-half times what it was in preindustrial times. Although these measurements clearly show the accelerating accumulation of these gases due to human activities, the effects they may have on planetary climate remain highly controversial.

The atmospheric models upon which the global warming theory is based are quite crude, according to Richard S. Lindzen, an MIT climatologist and an outspoken critic of the more dire forecasts. He argues that the effects on global warming from increased levels of carbon dioxide and other greenhouse gases are probably widely exaggerated. He points out that the data used to indicate that global warming may already be occurring are also highly suspect, and that the public outcry—fueled by statements from well-meaning policy makers who have little understanding of the weaknesses of atmospheric modeling—may lead to costly actions that could prove to be unnecessary.

In spite of such doubts about the validity of global warming fears, it is likely that some steps will be taken to reduce greenhouse gas emissions. Environmental issues with worldwide implications like this one, as well as those that have triggered transborder disputes (acid rain and ocean dumping), have become the subject of a series of international forums.

In 1972, after an international conference in Stockholm, Sweden, aimed at exploring options for international cooperation, a United Nations Environmental Program (UNEP) was established as an international organization. Several other international environmental meetings ensued.

The trend toward world-scale agreements is perhaps best exemplified by actions that are being taken to deal with ozone depletion. Worldwide environmentalists began to call for rapid international action when scientists discovered holes over the polar regions in the ozone layer atop the earth's atmosphere. This thin ozone blanket helps filter out harmful ultraviolet rays from the sun. Experiments demonstrated that the problem was related to the release of CFCs used in some aerosols, refrigerants, and solvents common to all industrialized nations. Because of the potential health dangers, an international meeting was held in Vienna in 1985 resulting in the Vienna Convention for the Protection of the Ozone Layer. This was followed in 1989 with another international conference in Montreal, Canada, that produced the Montreal Protocol, calling for the gradual phaseout of ozone-depleting substances. The Montreal Protocol also required periodic scientific assessments, which subsequently have revealed much more rapid ozone depletion spreading over a much wider region of the earth than had earlier been predicted. As a result, amendments were added in 1990 to toughen the Montreal Protocol. Increased environmental pressure in 1992 led the United States to speed up its own timetable for the phaseout of CFCs.

In June 1992, the United Nations sponsored the "United Nations Conference on Environment and Development" (UNCED), which took place in Rio de Janeiro, Brazil. The aim of the conference was to discuss and reach an agreement about the directions for global action. This Earth Summit developed a number of proposals for new international accords. The unprecedented participation of heads of state from countries large and small set in motion worldwide discourse on such issues as the assistance the industrialized world can give to poorer nations to help them follow an environmentally sustainable course of development, as well as methods for cleaning up heavily polluting industries of the former Soviet bloc.

The conference produced several important steps in the direction of global environmental cooperation. The delegates agreed on a Global Warming Convention that recommends curbing emissions of the important greenhouse gases that could contribute to raising average temperatures on the earth. This agreement commits each nation to work on cutting its own emissions, but, at the insistence of the United States, it does not set actual targets. The conference also produced Agenda 21, an 800-page document outlining the process of cleanup for the global environment and encouraging environmentally sound development. The conference also ratified a Declaration on Environment and Development that lays out 27 broad principles for setting environmental policy to balance development and ecology, while providing developing nations with special priorities for increasing their standard of living. A Statement on Forest Principles reached at the conference requires that countries assess the impact of development on their forests and then take steps, individually and in collaboration with other nations, to minimize any damage.

The global warming issue, which involves much greater scientific uncertainty, has been marked by far more divisiveness in the world community. While the agreement that came out of the Earth Summit in Rio does require each nation to set targets for limiting greenhouse gas emissions, it does not prescribe how the targets should be reached, and there is no enforcement power. This was a weaker pact than had been urged by European nations and Japan. Their negotiators wanted to limit emissions formally in the industrialized countries by the year 2000 to 1990 levels. The United States forced the compromise, because it is by far the largest burner of fossil fuels while Japan and some

European countries, such as France, use a much larger percentage of nuclear power. Environmentalists challenged the U.S. government's position, claiming that some studies showed that reducing such emissions through more efficient use of energy would actually save money and create jobs, rather than leading to higher costs and a loss of jobs as government officials claimed.

The argument of those who demand immediate response in spite of valid questions about the scientific basis for global warming is that the impact would be so devastating to civilization that we can't afford *not* to take prompt preventive action. By the time the data becomes clear, they reason, the Arctic and Antarctic ice sheets may already be melting while desertification is creeping steadily outward from equatorial regions into temperate zones. Furthermore, by then the warming trend may have become irreversible so that actions taken in later times could prove ineffective.

To slow the buildup of carbon dioxide, efforts are being made to prevent further worldwide deforestation, particularly of dense rainforests, because this greenery helps store carbon dioxide which it has absorbed from the atmosphere. This objective was supported by one of the agreements coming out of the Earth Summit. Some reforestation may also be tried as an offset to the buildup of greenhouse gases. This will help absorb carbon dioxide as the forests develop, but there are questions about the effectiveness of this approach in the long run. Costs of reforestation go up sharply with the amount of offset, and once forests mature they contribute little to net carbon dioxide absorption. Eventually, unless lumber is harvested and preserved in structures, carbon dioxide will be returned to the ecosystem either as trees rot and decay or are burned as fuel.

Aside from the dispute over stabilizing greenhouse gas emissions at 1990 levels, some independent steps are already being taken to reduce carbon dioxide levels. Denmark, for example, has vowed to cut its carbon dioxide output to half the 1988 level by the year 2030, while former West Germany aims to cut its emissions to 75 percent of the 1987 level by 2005. Meanwhile, in the United States, two major California utilities, the Los Angeles Department of Water and Power and SCE of Los Angeles, announced in May 1991 that they were setting their own targets of a 20-percent reduction in carbon dioxide emissions from 1988–89 levels over the next decade.

Although there are critics who see such proclamations as little more than "pie in the sky" without specific strategies spelling out just how such major reductions could be achieved in practice, the trend toward voluntary action seems to be gaining momentum. More and more executives appear to be concluding that it makes good business sense to have customers and potential customers perceive of their organizations as part of the vanguard of those working to preserve the environment. It seems doubtful, however, that such individual actions will be sufficient to deal adequately with a growing array of global environmental concerns.

Instead it appears that international forums like the Earth Summit will be essential, not just to deal with potential transborder disputes and to minimize the environmental degradation resulting from industrialization and development, but also to ensure that spreading civilization does not eventually so degrade our planet that it becomes much less suitable for human life, with increasing health problems and mortality rates. Some predict that the set of accords that came out of the 1992 meeting in Brazil will do for the international community what the United Nations Charter did in 1945. The UN Charter set out to regulate, within stated principles, acceptable actions between nations. The various conventions and other agreements signed at the Earth Summit help to codify principles for regulating actions between human beings and nature. Since the construction industry is at the forefront of this interaction, the parameters, policies, and legal perspectives embedded in the agreements could have great significance for the industry's future.

CHAPTER 4

TECHNOLOGICAL CHALLENGES AND OPPORTUNITIES

In Chapter 3 we explored some of the reasons why traditional economic forces fail to protect the environment. This chapter expands on that theme to explain why traditional regulatory policies have not proven effective in providing the impetus needed for the development of new, environmentally friendly technologies. Command-and-control regulations focus on the control, treatment, and disposal of wastes once they are generated. By setting specific targets or specifying the methods or equipment used for treatment or control, this approach tends to freeze environmental technology at levels just adequate to achieve compliance. There is little incentive for vendors to improve their offerings if industry is not forced to improve environmental performance beyond defined requirements. Perhaps even more critical, there is little inducement for industry to adjust its basic processes to reduce or eliminate waste and pollution in the first place or to increase reuse and recycling. Policy makers are now considering new market-incentive methods that shift emphasis away from end-of-pipe technologies and instead provide economic incentives for environmental protection. Some market-based policies would encourage industry to go beyond just controlling and treating pollution and wastes, toward eliminating or minimizing them by means of source-control technologies that involve the full range of industrial operations. Other incentive mechanisms would encourage consumers to change their behavior, cutting back on waste and allowing reuse or recycling when appropriate.

In order to reach society's environmental goals, we will need to make fundamental changes in the way economic activity and development is carried out. While making better use of existing technologies might slow environmental decline, it will not be sufficient to deal with rapidly growing problems stemming from population expansion and increased urbanization and industrialization around the world. Coping with these challenges will require advances not only in the technologies for standard environmental remediation and treatment, but also in technological alternatives that will significantly reduce or avoid environmental contamination across virtually every industry, including construction.

Although construction firms may not be involved in developing these new technologies, they will be affected in both the methods they use and the waste they create, as well as the markets they can expect to face in the coming decade. Construction firms will be instrumental in getting many of these new technologies into the marketplace and will have to adapt their own methods and operations as technologies change.

This chapter will discuss the effects of both traditional and market-incentive regulatory policies on the direction and pace of technological development. Two potential growth areas of particular market interest to the construction industry—waste-management systems and energy generation—will be treated in some detail. Finally, there are currently institutional and other barriers to innovation in environmental technology, often built into the system set up ostensibly to protect the environment. These will be discussed, along with a few of the efforts being made to reduce them.

REGULATION DRIVES THE ENVIRONMENTAL MARKET

The concept of sustainable development was introduced in earlier chapters as part of a new paradigm recognizing that economic progress will be stifled unless industry develops new operating modes that preserve environmental quality for both the present and future generations. Alternate technologies are required not just for disposing of waste and pollution, but, even more important, for helping to minimize or even eliminate their generation at the source. The system of rewards and penalties that drives individual decision making, for both producers and consumers, needs to be skewed to favor sustainable development and use of innovative, "environmentally friendly" technologies.

Can these needed technologies be expected to evolve readily within traditional markets? Powerful incentives for technological progress are built into the economic system for other sectors, but not, unfortunately, for dealing with environmental problems. Indeed, there are many instances in which advancing technology contributes to environmental decline by creating harmful by-products or vastly speeding up the exploitation of natural resources. Even with the system of regulation that has developed over the past couple of decades, the current economic arrangement actually continues to *encourage* environmental abuse in many cases. Since the costs to society for environmental damage are not naturally assessed to those creating the problems, firms that reap the greatest profits are often those doing the most effective job of avoiding, or externalizing, such costs, while those that strive to be good corporate citizens may be penalized for their altruism.

Consider a company that decides independently to invest in the development of advanced pollution-control and waste-management technology to clean up its operations. Even if the firm hopes to capitalize on this new environmental technology by selling it to others, it is unlikely that there will be much of a market unless other firms are required by regulators to improve their environmental performance to match this firm's, or unless the prevailing regulatory and economic system rewards the green firm (by lowering its taxes or by creating a market for its products) or sanctions the evading firms (by raising taxes or denying markets to their products). Its competitors may concentrate instead on investing in productivity-enhancing equipment and in new product R&D. After a time, the green firm might find that the costs of its environmental developments are eating into profits, while it does not have enough capital to keep up with competitors in either its factories or the market.

If a company in the environmental marketplace develops a new type of flue-gas scrubber that does a more effective job of removing harmful chemicals from emissions, it can't expect to make many sales unless regulatory authorities target these pollutants, requiring industrial plants to remove them from emissions. Regulatory pressure would then encourage plants to install this more advanced equipment. Thus, unlike normal markets, demand for technology in this sector must initially be triggered by regulation. Once the demand is in place, then market forces will begin to shape the size and competitiveness of the marketplace, the types of technology offered, and the pace of R&D.

Taking the example of a company exercising environmental leadership a step further, if there were regulations requiring further reductions in sulfur dioxide and the new system is judged to be the best available for this purpose, then this advanced technology might be specified for all new plants and perhaps even designated to replace existing scrubbers.

Should this company then continue its innovative work, hoping to find other effective solutions to environmental problems? It might envision developing and marketing a better method for cleaning up fumes resulting from spray-painting operations. There would be little incentive to do the necessary R&D without some certainty that the authorities would tighten standards for all those with spray-painting facilities. If management believed there was little chance that plants that had already installed existing, less effective systems would be forced to retrofit, then it might conclude that a market consisting only of new plants with spray-paint units would be insufficient to warrant further investment in R&D. Meanwhile, the factories that had installed cleanup equipment would resist moving to new methods of either controlling or eliminating emissions because they already had invested their capital in systems that met applicable regulations.

In fact, because the marketplace is so dependent on the action of regulators, if policy makers shift direction in the future the market could dry up completely. Taking the case of flue-gas emissions, policy changes might lead to regulators pressing industrial plants to move to cleaner fuels instead of ordering them to install better scrubbers. Any company developing new scrubbing technology would lose its total investment.

The type of regulation, the timetable set for compliance, the degree of enforcement, and the consistency of environmental policy all serve as critical factors in determining what kind of technology is developed and the marketplace for it. Commercial success thus depends significantly on the whims of regulators rather than the effectiveness of new technology or its value in improving environmental performance. Risk takers in the environmental arena must consider, in addition to the competitive factors that must be faced in any marketplace, that political changes and shifts in the regulatory framework could eliminate a potential market overnight.

HOW TRADITIONAL TYPES OF REGULATION IMPACT TECHNOLOGICAL DEVELOPMENT

The attempts by the U.S. government to control air emissions from the nearly 27,000 major and hundreds of thousands of minor stationary sources of air pollution provides some simple illustrations of the impact of different types of regulation on the development of technology and the shaping of the marketplace. Stationary sources are responsible for the gamut of airborne waste problems from particulates and acid rain to greenhouse gas emission problems. The primary target of this type of regulation is, of course, the energy-generation industry.

In 1970, the newly formed EPA's first attempt at controlling the emission of airborne waste was to establish the National Ambient Air Quality Standards. These **performance standards** represented the maximum permissible concentrations of some common air pollutants in specific geographic regions (called air quality control regions) that share common air quality concerns. Primary standards were to be those protecting human health; secondary standards were to be established if the health-based standards were insufficient to protect exposed materials, agricultural products, forests, or other nonhealth values.

This type of regulation had several effects—some positive but also some negative—on the market and technological development . First, by measuring air quality in a region rather than in specific plants, it becomes a disincentive for an individual firm to independently decrease its pollution. In some cases, it proved to be more cost-effective simply to build taller smokestacks in order to dilute the emissions over a wider area. Unless specifically directed by the regulating agency to do so, many smaller firms chose to delay action since it was the largest firms who were initially singled out. Second, by 1975 it had become clear that, despite some major gains in air quality, many areas had not met and would not meet the ambient standards by the statutory deadlines. The EPA eventually did extend the compliance deadline to 1982.

Under the 1977 amendments to the Clean Air Act, along with extending the deadline, Congress gave the EPA the power to set binding emission standards for all new sources of common air pollutants. These were known as new source performance standards (NSPSs) and had the important characteristic of being based on the "best technological system of continuous emission reduction" (in other words, these were **technology-based standards,** determined by the state-of-the-art in pollution control at the time the standards were set).

Over time, Congress decided that in order for out-of-compliance air quality control regions (called nonattainment areas) to meet their applicable standard, new sources wishing to locate in these regions had to install technologies consistent with the "lowest achievable emissions rate" (or LAER). Furthermore, in some areas where stricter air quality is called for than the NAASQs, major new sources were required to install the "best available control technologies" (or BACT).

These developments authorized the EPA to establish technology standards on all new plants (usually, regulations do not explicitly specify the technology, but establish standards based on a particular technology). States were then given individual control as to how they would achieve NAASQ compliance with existing plants within their own borders. States were required to submit a state implementation plan (or SIP) to the EPA to describe how they would achieve the standard.

This transfer of responsibility to the states had some potentially detrimental effects on the uniformity of national standards. Suppose pollution in one jurisdiction could be transported by prevailing winds to another jurisdiction; then the first region could impose weak controls on its local sources, thereby enhancing its economic position without suffering from the associated pollution. And vice versa, the region receiving the pollution might be quite stringent in controlling sources within its jurisdiction, only to suffer from the pollution imported from the neighboring region. The EPA attempted to offset such inequities.

The cumulative effect of these actions has been to slow the development of new technologies and to create perverse economic incentives for air polluters. Companies wishing to develop technologies that could go beyond the emission standard would find no market unless the standard was subsequently adjusted to take the new technology into account. Likewise, existing industries situated in regions that were attaining the NAASQs faced little incentive to reduce their pollution levels, especially if competitors located within nonattainment zones had to achieve stricter standards. This provided them with a cost advantage in their market. In both cases, the regulation, as designed, stifled the incentive for innovation.

Thus, while policy makers recognize that regulation is necessary to achieve environmental goals in the marketplace, they must find mechanisms that are less restrictive on innovative approaches when fundamental changes are crucial. An alternative approach to fostering technological progress toward achieving environmental goals such as cleaner air might be through tying regulatory goals to market incentives and then allowing the power of marketplace competition to lead to a variety of innovative technologies for meeting the targets.

MARKET-INCENTIVE REGULATIONS

Market incentive regulations can be divided into two categories: indirect and direct.

- Indirect market-incentives do not assign a value to pollution directly, but rather affect the status of available options for dealing with that pollution. One example of this type of incentive is the *land disposal restriction* under the 1986 Hazardous and Solid Waste Amendments to the Resource Conservation and Recovery Act. These restrictions established a mandatory retirement age for the nation's landfills and created rigid requirements for the creation of new ones. As a result, the EPA estimates that 80 percent of existing landfills will be closed by 2010. As landfill space diminishes, costs for this and other forms of disposal are increasing. For example, landfill costs in New Jersey have increased from $11.88 per ton in 1987 to as much as $42.50 per ton in 1990.

    In the broader sense, the general proliferation of increasingly complex and costly regulations is an indirect market incentive to minimize pollution. Each additional regulatory burden increases the costs associated with the storage, treatment, and disposal of toxic wastes. As an example, RCRA, CERCLA, and the Clean Air Act have all been amended within the past 5 years. Each amendment has created more, not less, regulatory control over industrial activities. The 1990 Clean Air Act alone has been cited as the most expensive piece of environmental legislation ever passed, with an estimated burden on industry of $25 billion per year.
- Direct market incentives apply a direct economic cost or value to the polluting substance. The recent Project 88 initiative discussed in Chapter 3 outlines three ways to create these costs: (1) *Pollution charges* impose a fee or tax for every unit of pollutant that is discharged into the environment. (2) Under a *tradable-permit system*, the allowable overall level of pollution is established and then allotted in the form of permits among firms; firms that keep emission levels below the allotted amount may sell or lease their surplus permits to other firms who have been unsuccessful in attaining the standard. (3) Under a *deposit-refund system*, surcharges are paid when customers purchase potentially polluting products. The surcharge is returned when the consumer returns the product to an approved recycling or disposal facility. This system has been proposed for materials such as batteries and oils.

The primary economic objective of such regulations is to force all industries to share the burden of lowering overall emissions. Since the cost of complying with a given emission standard for inefficient old industries can be as much as 100 times that for efficient new industries, market-incentive regulations, particularly tradable permits, allow low-polluting companies to sell their credits to high-polluting companies at a rate set by the market systems. By necessity, the established price will be less than the total compliance cost for the inefficient producer, but it will be greater than the total compliance cost for the efficient producer. If this were not so, there would be no incentive to trade. As a result, companies have an incentive to minimize pollution as much as possible through technological development as long as a market exists for the credits. Different facilities might exceed the standards by installing more efficient energy-generation systems, by changing fuels (from high-sulfur soft coal to low-sulfur hard coal, or from oil to natural gas), or by doing a better job of cleaning up effluents. Because such rights might be worth millions of dollars, capital becomes available to pay for new, more environmentally sound technology. Meanwhile, some older plants with higher emissions might be kept in operation for a limited period while newer, less-polluting facilities are being built. The result might be that the costs of new systems can be covered at least partially by the pollution reduction achieved. This approach allows progress toward lower emissions without so much disruption of business and industry, not to mention the communities where older plants are located and workers who would probably be laid off if their plants abruptly shut down. The United States is now experimenting with applying the tradable-permit system for controlling air pollution via the 1990 amendments to the Clean Air Act. Many nettlesome problems face regulators in managing a tradable-permit system, especially as a result of regional differences. They may wish to avoid having permits purchased so that factories can continue to exceed standards in heavily polluted areas, but at the same time they don't want to encourage heavy polluters to migrate to clean areas.

As the technologies for reducing emissions advance, regulators might reduce the total level of allowable emissions step-by-step, thus steadily providing new capital for the next round of installations of less polluting systems. This is an approach that does encourage new technology, but it also sets in place slowdown mechanisms geared somewhat to the life-cycle of capital equipment. This serves to keep progress from being *so* rapid that it causes major disruptions to the business world and the society. Some activists consider the accelerating pace of environmental decline to be so threatening to ecosystems, however, that they are demanding more rapid change regardless of any disruption this might induce. Environmental groups have even suggested buying permits themselves from firms that have reduced their pollution levels and then not using them. To insure that transaction costs are kept low, the EPA has authorized the Chicago Mercantile Exchange to set up a trading board for pollution credits.

ENVIRONMENTAL PROGRESS: SALIENCE OF TRADE-OFFS

The environmental sector is permeated with the need to find suitable trade-offs. Practicality requires that the desire to cut out all pollution be tempered by the reality that, if industrial and other activities are to be allowed, then some pollution is inevitable. Environmental goals may seem clear and simple: clean air; clean water; conservation of forests, arable land, and wetlands; and efficient utilization of natural resources such as hydrocarbons, metals, and mineral deposits. Yet achieving these goals is never a black-and-white matter. The worldwide clamor for better environmental protection is creating demand for improved methods and systems, but economic and political realities make progress difficult. Some new technologies are helpful in coping with environmental decline, but others are making the problems even more severe. Thus, as regulators attempt to spur the development of alternate technologies, it is essential that they take a macroeconomic view of the impact of the solutions being sought.

To take a simple example: A certain quantity of BTUs are required to make a pound of glass. If the glass is buried, 100 percent of that initial energy is lost. If the glass is burned, perhaps 15 percent of the original BTUs will be recovered. But if the glass is recycled, up to 80 percent of the energy required to make new glass containers to replace those that had previously been thrown away might be saved. In choosing an option like this, however, it is also important to consider how efficient a system can be set up for collecting used glass, transporting it to recycling facilities, cleaning it, and converting it to usable forms. The energy and cost savings from recycling the material could easily be eaten up in this elaborate process. The importance of the interplay between the industrialized and natural worlds must be recognized as new technology is developed and applied.

TECHNOLOGIES FOLLOW DIRECTIONS SET BY POLICIES

Sound and well-balanced environmental strategies must address these practical issues while fostering innovative technologies that can help offer solutions. Policy makers might structure incentives and penalties to favor waste minimization over waste management that will insure that our industries are competitive in the long run but not at a pace that will limit growth in the short run. In the case of automobiles, rules could be skewed toward encouraging the development of cleaner—or even completely nonpolluting—engines, rather than the use of catalytic converters. In the hazardous waste area, rising costs for containment, transport, and disposal of toxics and other harmful substances are likely to lead to more processes that avoid creating them in the first place but do not force business into bankruptcy. Where the problem is serious enough, as has proven to be the case with CFCs and their impact on the ozone layer, an outright ban is the only way to ensure that substitute technologies are quickly put to work. In systems for production, some policies might favor processes that generate fewer pollutants (by implementing a pollution tax or fee system), while others might encourage conservation in consumption (such as with carbon and fuel taxes).

To ensure compliance, it is essential that, even though environmental rules may have to be toughened at times, regulators must set reasonable timetables if they hope to get cooperation and compliance from industry. Environmental goals also need to be integrated with other policy issues, such as economic development, rather than being formulated and enforced in isolation. Policy makers need to develop long-range vision so that business and industry can plan effectively to meet future requirements and also so that they can recognize emerging market opportunities within the environmental sector.

Even though there is a clear need for new or modified technologies, experience has shown that encouraging their development and then getting them widely accepted will be a continuing struggle. While some policy makers would like to shift environmental strategies from an emphasis on waste management and disposal to source control, there are practical limits on the pace with which such shifts can be accomplished. Past practices have a powerful momentum, and often the best that can be achieved is a gradual transition to "greener" practices. Some of the difficulties policy makers face in making progress will emerge from a brief review of technologies being applied in the environmental area, along with discussion of trends for the future.

Because of the command-and-control approach that has dominated the environmental area in the past, the technologies that have emerged have tended to be regulation-driven. In the future, policy makers hope that more market-driven technologies will emerge, sending the right signals to the marketplace. These technologies would respond to market demands for greater effectiveness in eliminating wastes, or in rendering them environmentally benign. Market pressures might also achieve better results at less cost. Existing plants have already invested in equipment to clean up their emissions and discharges to meet regulations, so it would not be fair to force them to immediately scrap these systems. There is also a need for interim technologies that address the problems of cleaning up the mess of the past. The markets that develop for these technologies may grow rapidly, but they will last for only a limited duration because, even though extensive cleanups will be needed at many thousands of sites, eventually the task will be completed. Meanwhile, more pressure will be applied to encourage industry to clean up wastes as they are created rather than dealing with disposal afterwards. And, until technologies exist to achieve source control, regulators will have to be satisfied with tightening restrictions on emissions and discharges while trying to encourage more ideal solutions.

What follows is a discussion of two technology markets that are growing rapidly in spite of the lack of market-incentive regulation. Command-and-control regulations have developed both the waste-management (hazardous and solid) and energy markets for developers of new technologies and the subsequent market entry by the construction industry.

HAZARDOUS WASTE

Perhaps the most pressing environmental problems are those dealing with the past disposal of toxic materials and other hazardous wastes. There are potentially hundreds of thousands of existing hazardous waste sites that must be cleaned up; many river and harbor bottoms that have been contaminated by PCBs; and other legacies of the past, including asbestos insulation and radioactive wastes from aging nuclear power plants. Aside from dump sites that must be cleaned up, there are huge numbers of potentially leaky underground tanks holding hazardous fluids. The booming markets that have been created to meet these cleanup needs have been a magnet for many construction firms, and the growth potential is expected to be strong for years to come.

Under the Superfund (CERCLA) program, 26,000 sites have been listed in a Hazardous Ranking System, with about 1,250 of them on the National Priority List. A more comprehensive inventory by the EPA would swell the numbers to nearly 370,000 sites, the GAO estimates; and this does not include many DOE, DOD, state and private cleanup sites, plus myriad leaking underground storage tanks, some at chemical plants but also including gasoline tanks at service stations. And these regulations do not apply to millions of home heating oil tanks. Because of the staggering dimensions of the existing problem, there has been a burgeoning market for technology suitable for cleaning up hazardous wastes.

Early efforts at cleaning up hazardous waste sites were ineffective. One approach was simply capping and containing, grading, and revegetation, with the intention of preventing further migration of harmful materials into the environment. This was little more than leaving a ticking time bomb for future generations. Another approach was the removal of waste and contaminated soils from the hazardous waste site to a landfill for burial. Later these landfills themselves were found to be leaking, and thus have been added to the National Priority List (NPL). This approach turned out to be little more than moving the problem from one place to another. Eventually these methods were succeeded predominantly by incineration, which, in effect, changes the chemical structure of most hazardous materials as they are converted from liquids and solids to gases. Some constituents are oxidized to less noxious forms but others remain chemically unchanged, and a few break down to produce potentially more hazardous by-products. There is a continuing effort to find treatment methods that could transform the hazardous materials into innocuous forms.

Remediation technologies must address both soil and any groundwater contamination using one of three basic approaches: in-situ, prepared-bed, or in-tank reactor methods.

In-Situ Systems

These involve treating tainted soils or groundwater right at the site where they lie, without requiring any excavation. This type of treatment involves serious complications due to the difficulty in reaching all levels of contamination and then verifying that remediation efforts have been complete. Needless to say, the cost savings of this type of remediation make it a very attractive option.

Prepared-Bed Systems

There are two versions of this approach. In one, the contaminated soil is removed from its original site to an area that has been designed to prevent transport of hazardous materials from the site and, in some cases, to enhance treatment. Another method, useful only when contaminated groundwater is not present, is to transport contaminated material to a holding area while the original site is prepared for treatment. Then the soil is returned and treated.

In-Tank Systems

This involves removing contaminated soil or groundwater for treatment in an enclosed reactor (using batch, complete mix, or plug-flow approaches). When this approach is used for groundwater it is commonly called "pump and treat."

There are also a wide variety of treatment technologies for hazardous materials, which can be broken down into five basic categories: thermal treatment, solidification/stabilization, physical separation, chemical treatment, and biodegradation.

THERMAL TREATMENT

There are both high-temperature and low-temperature forms of this type of treatment. The high-temperature approach operates at 2,500–3,000 degrees F, destroying and breaking down hazardous wastes into other compounds. Most technologies require that contaminated material be either continuously fed or batch-loaded into a reaction chamber under controlled conditions. Notable exceptions include high-temperature in-situ vitrification and low-temperature thermal treatment, where the work can be done while contaminated material remains in place rather than being loaded into reaction chambers. Complete combustion results in harmless gases and ash, but incomplete combustion can result in by-products that may be toxic. The EPA requires commercial incinerators to destroy the main hazardous constituents with a destruction efficiency of 99.9999 percent. This does not include noncombustibles, such as inorganics. Some also argue that, even though high efficiency may be demonstrated in a pilot project, this does not prove that the working plant will achieve the desired level. If a plant handles very high volumes of waste, even a tiny percentage of unvolatized material could put significant quantities of hazardous waste into the atmosphere.

There are four basic types of high-temperature treatment: incineration, pyrolysis, wet oxidation, and vitrification. **Incineration** destroys combustible components at temperatures higher than 2,200 degrees F. **Pyrolysis** decomposes organic materials in an oxygen-deficient atmosphere. **Wet oxidation** destroys organics at high temperature and pressure in a water solution or suspension. **Vitrification** destroys organics and immobilizes inorganics within melted glass.

High-temperature treatment is a more expensive approach than low-temperature methods, because the waste material seldom has a BTU content high enough to sustain combustion; therefore an auxiliary fuel, such as oil, gas, coal, or electricity is required for complete destruction of the waste. But high-temperature methods are the most suitable for organic waste; so, in spite of the costs, they are preferred by the EPA for highly toxic compounds such as dioxin and PCBs. Inorganics are condensed in the process and can be removed in the ash. An important side benefit of these methods is that long-term liability can be removed by destroying all harmful materials. But to ensure that destruction is complete, secondary treatment is also required. Flue gases must be treated, usually by a series of scrubbers and filters, before they can be emitted to the atmosphere. Complete thermal treatment, therefore, also requires the operator to deal with combustion ash (fly ash), scrubber liquid, and filter dust. These are all considered hazardous waste and, unless delisted, must be treated again or disposed of properly. Because of this additional waste and questions about the long-term effects from emissions from incineration plants, there has been intense public opposition to them and consequent difficulties in obtaining permits.

In the United States, thermal treatment may be applied either at a fixed incineration site or within a mobile unit erected wherever the contamination is located. In Europe, incinerators on ships have been burning hazardous waste out on the ocean since 1969. One benefit of burning at sea is that the main air pollutant emitted by burning chlorinated materials—chlorine—is absorbed by seawater as hydrochloric acid droplets are formed due to the water vapor in the air above the ocean surface. This method has not gained acceptance in the United States and is also losing popularity abroad. The main objection is the possibility that a ship laden with hazardous materials might sink, releasing toxic materials into the ocean. Another objection is that harmful gases other than chlorine compounds are released into the atmosphere. There are also local groups opposed to handling hazardous substances at ports where the ships would be loaded.

The most common high-temperature treatment technology is the **rotary kiln incinerator.** Contaminated soil is continuously fed into a reaction chamber—the rotary kiln—where intense heat vaporizes hazardous components. Any residual contaminants are destroyed by injecting emissions from the kiln into an afterburner.

Convenience and efficiency are offered by treating hazardous waste right where it is rather than having to move it. One innovative technology, **in-situ vitrification,** offers this capability. Electrodes are placed in the ground around contaminated soil and electric currents are passed between them. This melts the soil and even surrounding rock. Organic components, including PCBs, are destroyed by pyrolysis as temperatures climb as high as 4,000 degrees F. Inorganic contaminants are trapped within the vitrified, glassy mass. A vapor-collection system must be installed over the site to capture any organic or inorganic emissions. The process was developed by Battelle Pacific Northwest Laboratory under a grant from the DOE. Currently, the DOE has licensed only one company, Geosafe Corp., a subsidiary of Battelle in Kirkland, Washington, to apply this technology, but the system has been temporarily withdrawn from the market until safety problems have been solved. There is a tendency for the apparatus to explode.

Low-temperature thermal treatment avoids more costly incineration or pyrolysis by operating at levels of only 200 to 900 degrees F, separating organic contaminants from soils, sludges, and other solid media through evaporation. Heat is used to remove contaminants, but not to destroy them as in the high-temperature approach. These processes avoid chemical oxidation and other reactions so that no combustion by-products are formed. Organics are removed in a condensed, high-BTU liquid, which must then be destroyed in an incinerator with a proper permit. An innovation is this field is the X*TRAX system developed by Chemical Waste Management, Inc., of Oak Brook, Illinois, which performs low-temperature thermal treatment in a mobile reactor vessel.

Toxic Treatments, Inc., San Francisco, has developed another innovative in-situ process utilizing low-temperature thermal treatment. Injectors pump air (at

300 degrees F and 250 pounds per square inch [psi]) and steam (at 450 degrees F and 450 psi) through rotating cutter blades drilled into the contaminated site. Heat from the steam causes the organics to vaporize, and the combination of steam and air carry the stripped contaminants through a collection system above the blades up to the surface, where condensation removes them from the exhaust stream. The condensed water is then distilled so that the contaminants can be removed.

SOLIDIFICATION / STABILIZATION

These techniques do not destroy the hazardous materials, but rather reduce their mobility by facilitating chemical or physical actions. Implementing them generally requires extensive handling and mixing of the materials, limiting in-situ applications, but the work can all be done on site, in tanks or containers. These processes are relatively inexpensive in comparison to other technologies, but they are also less permanent.

Solidification generally results in a durable, monolithic block of material. Stabilization is accomplished by adding and mixing materials that limit the solubility or mobility of waste constituents even though the physical characteristics of the waste may be unchanged. The mass of treated waste has higher strength, lower permeability, and lower leachability than untreated materials.

These approaches are most suitable for inorganic waste containing heavy metals. Organic materials can create problems because they may interfere with the setting action or leach out of the treated waste over a period of time. The techniques can be quite useful as a short-term solution for emergency situations or at sites where space is limited. Waste can be converted into a more transportable form so it can be moved elsewhere for further processing.

There are drawbacks to solidification/stabilization methods, however. Considerable materials handling and controlled mixing is required, creating up to double the weight and volume of the original waste material. Contaminant leaching also occurs over the long term due to the porosity of the resultant solid mass.

Three commonly used methods involve traditional technologies. **Cement-based solidification** chemically or physically seals the waste into a matrix with a portland cement base. This neutralizes and seals acids and handles strong oxidizers. In addition, because cements are highly alkaline, they solidify many toxic metals by forming insoluble carbonates and hydroxides. In the case of organics, however, this method is generally considered unacceptable. **Pozzolanic stabilization** binds wastes in a lime and siliceous matrix in a process similar to that used with portland cement. **Thermoplastic binding** seals wastes in a matrix such as asphalt and is particularly useful for materials containing limited concentrations of petroleum oils. There can be a problem, though, if the waste also contains organic solvents, strong oxidizers, or thermally unstable materials, because these tend to break down the matrix.

Three innovative technologies have recently been added to the low-temperature treatment category. **Glassification** combines a waste stream with molten glass, resulting in a solid, glassy residue resembling that produced by in-situ vitrification. **Ion exchange** removes contaminant ions from the waste stream, substituting instead innocuous ions from a binding material, typically clay. **Microencapsulation** is a molecular-level treatment in which calcium-alumino-silicate compounds solidify, fixate, and encapsulate hazardous wastes.

PHYSICAL SEPARATION

In these techniques, hazardous components are separated from the carrier soil and from each other through methods such as volatilization, adsorption, extraction, or filtration, without altering their chemical structure. In-situ vacuum/vapor extraction removes volatile organic compounds (VOCs) from the soil by applying a vacuum through a production well, forcing the compounds to diffuse into the well. In a similar system, air is pumped into an injection well and VOCs are withdrawn along with the exhaust air in extraction wells. This treatment process can also be carried out after waste has been moved (ex-situ). Soil aeration is a process in which soils are excavated and aerated in a mill or drum, causing VOCs to volatize so they can subsequently be collected and treated.

Soil Washing. This method extracts contaminants from excavated material using a liquid medium such as water, organic solvents, water and chelating agents, water and surfactants, acids, or bases. **Soil flushing** is applied while the soil is left in place using an injection/recirculation system. In both cases the contaminants are removed from the washing solution using a conventional treatment system.

Chemical Extraction. These processes serve to separate contaminated soil and sludge into components (phase fractions): organics, water, and particulate solids. Solvent extraction methods include B.E.S.T., using secondary or tertiary amine as the solvent, and critical fluid, which makes use of liquified gases such as carbon dioxide or propane at high pressure.

Freeze Separation. This is an innovative technology developed by Freeze Technologies Corp., Raleigh, North Carolina, that physically separates out waste by freezing it. The techniques are based on the principle that the natural ice crystals that form when water freezes exclude contaminants from the matrix, allowing them to be collected. In the freeze-crystallization process, refrigerant is injected into excavated soil, and the resulting temperature drop turns liquids into solids. Crystals of solute and solvent are separated from each other by gravity and then are melted in a heat pump.

Electro-Osmosis. This method is the basis for remediating contaminated sites in a process under development by researchers at MIT. The principle involved here is that water can be made to flow through saturated clay by applying an electric field. Two electrodes are spaced apart within a saturated zone of contaminated soil, and a current is induced. Clean purge water is added at the anode (the positive electrode), and contaminated groundwater then flows toward the cathode, where it can be removed and treated. The technique is still experimental, but it offers advantages over traditional groundwater-pumping methods because the flow direction and distribution is very uniform and controllable.

Physical separation techniques such as these must always be used in series with other treatment techniques because they are primarily useful for separating and condensing hazardous contaminants. The resultant waste streams must then be treated for detoxification or destruction before final disposal.

CHEMICAL TREATMENT

These techniques destroy or detoxify hazardous components through the use of chemical oxidation or reduction reactions. **Oxidation** reactions are generally applied to waste streams contaminated with organics because heavy metals (with the exception of arsenic) are more mobile at higher oxidation states. Contaminated soils are loaded into a reactor vessel and oxidizing agents, such as ozone or hydrogen peroxide, are added. It would be difficult to control applications where the soils are left in place (in-situ) because soil particles are non-uniform. Phenols, aldehydes, and certain organic compounds containing sulfur are highly reactive while other organics, such as halogenated hydrocarbons and benzene, are virtually impossible to break down using this method. Chemical **reduction** of soil contaminants is not as widely applicable, but soils contaminated with chlorinated hydrocarbons and some heavy metals are receptive to reducing agents.

Chemical treatment generally destroys contaminants completely, but in a few cases where only partial degradation takes place, the resultant compounds could be even more toxic than those in the original material.

An innovative system developed by Ultrox International, Santa Ana, California, makes use of ultraviolet radiation along with the oxidizing agents, ozone and hydrogen peroxide, to destroy organic compounds, especially chlorinated hydrocarbons, in water.

BIOREMEDIATION

Bacteria, fungi, and other microorganisms can be used to detoxify organic matter through a process of **biodegradation.** There are several methods of application, including composting, in-situ, solid phase, and slurry phase, which may take place either with oxygen (aerobic) or without oxygen (anaerobic). The processes are highly sensitive to environmental conditions such as temperature, acidity or alkalinity, illumination, and concentrations of the contaminants and microorganisms. Because of this sensitivity, the processes will only work under very controlled conditions, and remediation takes place very slowly. Some sites may take as long as ten years to remediate. Still, this approach promises to offer significant cost advantages, so intensive development work is under way in many laboratories.

Native organisms are generally extracted from the contaminated site and then tested to determine which types are most applicable for the waste to be treated. The chosen specimens are colonized to increase populations greatly, and then they are reintroduced to the waste site. Researchers are using genetic engineering methods to develop more effective strains for particular types of waste, but this work is still experimental.

Bioremediation was successfully tested by the EPA following the Exxon *Valdez* oil-spill disaster in Alaska. Phosphate and nitrogen fertilizers were added to oil-contaminated shorelines to multiply by 30 to 100 times the populations of native oil-degrading microbes. These microbes "eat" the oil, producing carbon dioxide and water. Within only three weeks, the fertilized areas had shown dramatic decreases in the levels of oil contamination.

Some researchers claim that bioremediation is capable of destroying PCBs and dioxin. Extensive research on these techniques is being conducted at General Electric's (GE) Corporate Research and Development Laboratory in Schenectady, New York. GE is awaiting approval from the EPA to employ aerobic and anaerobic microbes to clean up PCB-contaminated Superfund sites in California, Massachusetts, and New York.

Although there are some drawbacks, the advantages of bioremediation can be dramatic. Costs can be far less than for other methods—perhaps $100 per ton versus $1,000 per ton for incineration, for example. Also, the technology can be applied in a variety of in-situ conditions: in soil, groundwater, lakes, or rivers. Unfortunately, it can be extremely difficult to verify complete detoxification of waste for in-situ applications. This approach also suffers from the long time periods required from initiation to completion of a cleanup.

Innovative technologies for treating hazardous waste have been slow to gain acceptance in the Superfund program, but a few successes may help to spur greater use. So far, only six cleanups have been completed using innovative approaches, although new technologies are either in use or have been specified for 37 percent of the Superfund site cleanup plans issued over the period from

1982 to 1989. Support for R&D efforts is increasing, and it is estimated that investments by both government and industrial firms will amount to several hundred million dollars in 1993.

It is also important to consider the advances that have been made in new technologies for the treatment of waste as they are being created. Since the initiation of significant environmental regulation in 1970, the method of choice for dealing with pollution has generally been the construction of expensive end-of-pipe technologies that deal with waste after it has been created. Such technologies include flue-gas desulfurization and stripping of gaseous waste (scrubbers); pH neutralization and solids separation of liquid waste; and incineration and landfilling of solid waste. Due to the traditionally segmented nature of environmental regulations (solid versus liquid versus gas), many companies have often altered the medium of their waste to whichever has the least stringent requirements. For example, scrubbers produce a liquid waste that contains the contaminants that were originally in the stack gases. That liquid could then be filtered or incinerated, which would create a solid waste containing some or all of the original gaseous contaminants. Innovations in end-of-pipe technologies have been occurring over the past two decades in response to the proliferation of environmental regulations. However, alterations in the process are now being considered an option for pollution prevention.

Recently, with the rising costs of disposing of hazardous waste both from an operational perspective (cost of running end-of-pipe technologies) and from a final disposal perspective, companies have begun to see it as increasingly beneficial to avoid the creation of waste in the first place. Thus was born the new perspective of a hierarchy of waste management practices: source reduction first, recycling second, treatment third, and disposal as a last resort. The primary initiative of this hierarchy, source reduction, can be achieved through five approaches: 1) change raw materials of production, 2) change production technology and equipment, 3) improve production operations and procedures, 4) recycle waste within the plant, and 5) redesign or reformulate end products. In the future, innovations in these types of hazardous-waste-reducing technologies can be expected to increase.

SOLID WASTE

Solid waste management is another area where an increasing sense of urgency is putting pressures on the technology base. With more than 2,000 of the nation's 6,000 landfills being forced to shut down by 1993 and thousands more due to close over the next few years while tough new regulations severely limit creation of any new landfill sites, new ways must be found to deal with mounting quantities of trash. Public pressures, particularly from NIMBY (not in my back yard) groups, are restricting some other options, such as new incinerators. A review of the primary technologies for the four major modes of solid waste management in the United States (landfilling, waste-to-energy, recycling and composting, and waste minimization) will help to indicate where progress is needed.

It is important to recognize that none of the technologies work in isolation. Rather they are all parts of an interrelated solid waste management system. We can visualize the approaches as consisting of a continuum—starting "upstream" with waste minimization and recycling and composting, and moving "downstream" toward incineration and landfilling. Downstream technologies must be capable of handling a wide variety of discards as different types of waste are taken out upstream.

Sanitary Landfill

The landfill is an unavoidable component of the system, since nonrecyclables, noncombustibles, and the residuals of waste processing all require disposal. They also play an important leveling role, since they can accommodate wide fluctuations in the waste stream most readily. They do have limited capacities, however, so that new ones must continually be sited.

Sanitary landfills are essentially earthworks constructions providing compaction and stabilization of trash. When properly designed and operated, they minimize volume while protecting the public from health and safety hazards. Federal regulations now require modern landfills to include many features for environmental protection. Groundwater is protected from leachate by impervious liners with layers of dense clay and synthetic materials, including a flexible-membrane liner with at least two feet of compacted soil beneath it. Leachate is a waterborne discharge from the landfill induced by the infiltration of rain and melting snow that may contain traces of hazardous materials such as organics and heavy metals. Any trapped leachate that forms is carried by pipes from the bottom of the landfill up to the surface for treatment and disposal.

Landfill operators compact waste in layers and cover it with soil daily to eliminate odors and to discourage scavengers such as gulls. Pipes sunk through the layers collect methane gas, formed as microbes cause anaerobic decomposition within the landfill. Although this gas is potentially explosive, it can also be useful as a fuel. A monitoring well dug near the landfill into any nearby aquifer checks for contamination due to leakage.

Several states now require even stiffer standards offering greater leachate protection because many authorities in leachate chemistry and liner engineering warn that the minimum federal standards are too lenient. Instead, these specialists recommend a double-composite liner system consisting of ten layers in four major segments: leachate collection system; primary composite liner; leak detection system; and, secondary composite liner.

Collection of methane, a colorless, odorless gas that is toxic in high concentrations and explosive in confined areas, is now required by law for landfills that generate enough of it to exceed lower explosion limits. Since these requirements have gone into effect, much more landfill-generated methane is being sold to gas users or used to generate power. In the past, collected gas was commonly vented and flared.

When a landfill is closed, the operator is required by regulations to install an impermeable layer on top to prevent infiltration into or emissions out of buried wastes. This "final cover" must be monitored carefully throughout a thirty-year "post-closure" period, with the landfill owner providing care and maintenance as well as financial assurance for any problems that might arise during that time. Facilities such as airports, golf courses, and even a wildlife refuge have been operated on top of such final cover layers.

Meeting the much more stringent requirements set up for sanitary landfills in recent years has made the cost for building and operating them escalate rapidly. A recent study suggested that an 80-acre, single-liner landfill in Michigan would cost about $125 million to build and operate, and this does not include the costs of interest, insurance, taxes, bonding, contingencies, corporate overhead, remediation, and profit. Break-even tipping charges would be $11.20 per ton for the owner, assuming 1,000 tons of trash per day for 320 days a year for 21 years, using a discount rate of 10 percent. Meeting even tougher standards as promulgated by the EPA in September 1991 and also installing double liners as several states now require would push these costs even higher.

Only 25 percent of U.S. landfills were estimated to have systems for controlling the migration of leachate into groundwater in 1988, and only 5 percent recovered methane and other gases produced by decomposition. Yet some 90 percent of the waste stream landfilled in the United States at the time was buried in these uncontrolled facilities. The process of biodegradation that is supposed to take place within landfills is also labeled largely a myth by Dr. William J. Rathje, a waste specialist at the University of Arizona. In laboratory tests, for example, newspapers are finely ground before they are biodegraded in a few weeks or months under ideal conditions. In reality, newspapers dumped into landfills are never ground into dust, and conditions are far from ideal. Many authorities view the indiscriminate dumping of mixed wastes as the most environmentally unsatisfactory mode for disposal.

Incineration

Until 1970, when Congress passed the amendments to the Clean Air Act, waste incineration was often carried out in mass burning facilities with minimal, if any, pollution controls. By the mid-1970s, due to rising oil prices, interest arose in burning trash to generate electricity. Government development authorities at the regional and state levels actively encouraged waste-to-energy projects, seeing the potential for exploiting a "new" energy source that was increasingly becoming a problem for overburdened landfills. Incinerators were appealing because they had the potential to link easily into the existing solid waste stream, providing a new source of energy while reducing trash volumes and destroying pathogens and toxic materials. But they also had drawbacks, and these have resulted in growing public opposition in more recent times. Improper operation can result in the emission of chemical toxins and traces of heavy metals. Costs of incinerators have risen steadily, and some have had reliability problems. Also, their constant demand for large amounts of refuse can impede efforts for more recycling. Still, even though energy prices have moderated, incineration remains an important part of the solid waste management mix.

Each facility is designed for a specific level of material throughput. Designers forecast the physical properties of the waste to estimate the thermochemical characteristics of the furnace, as well as the thermal capacity of the boiler that recovers energy in the form of heat. Waste is not an easy fuel to burn since it is typically abrasive, nonhomogeneous, and corrosive, so furnaces must be designed to withstand especially harsh combustion conditions and to minimize other problems peculiar to solid wastes.

There are two fundamental design parameters that can be used to describe waste combustion facilities: the fueling method and the equipment type or configuration.

The two dominant methods of fueling incinerators are mass burning and refuse-derived fuel (RDF) systems. **Mass burning** is the combustion of unprocessed, mixed municipal solid wastes. Only bulky and "white goods," such as mattresses and large appliances are separated out for possible landfilling or even salvage. **Refuse-derived fuel** systems are designed to remove noncombustibles, turning refuse into a more uniform fuel that permits better furnace and boiler operation. Both processes are in use for large-capacity (400–2,000 tons per day) and small-scale (usually less than 400 tons per day) facilities. Together they account for about 90 percent of installed and planned incineration capacity.

The type of furnace and its configuration are also important characteristics in municipal solid waste technology. Three component designs dominate the U.S. market: refractory-lined, waterwall, and rotary incinerators.

Refractory-lined furnaces enclose the fireball created during operation within a combustion chamber lined with heat resistant (refractory) materials. This heat shield also protects the metal outside shell and superstructure from warping that could be caused by rapid, wide temperature swings during start-up and shutdown. Similar linings are used in many other types of furnaces, such as those used for metal processing, glassmaking, or ceramics production.

Waterwall furnaces are more efficient because they are designed to provide heat recovery for energy production directly by cooling the firebox with heat-exchanger tubes. The waterwall is a water-filled tube lining around the combustion chamber that keeps the outside shell cool while directly absorbing heat from the fireball. This process also requires less air than refractory-lined furnaces to ensure that chemical reactions are complete. This means that costs can be reduced by downsizing many components, including expensive air pollution control equipment.

Rotary furnaces have the advantage of extensive turbulent mixing action on the waste inside a rotating combustor, so that they capture heat that would be lost in refractory-lined furnaces. One proprietary hybrid, called the O'Connor Combustor after the vendor that developed it, is becoming common in the United States. It uses a rotary, water-cooled combustor to volatilize and pyrolize waste in starved-air conditions, and it then completes incineration inside a waterwall-lined postcombustion chamber. Unfortunately, these furnaces are sometimes plagued by "excursions," in which combustion reaches abnormally high temperatures. This can lead to pollution-control problems due to excessive nitrogen oxide formation.

Controlling emissions into the atmosphere requires any combustion system to include pollution-control equipment. Municipal solid waste incinerators must abide by provisions of the Clean Air Act as well as regional and local emissions regulations. This requires devices for removing particulates plus additional equipment for removing acid gases from the flue-gas stream. Common types of particulate-removal devices include **electrostatic precipitators, fabric filters** (also called **baghouses**), and **granular filters.** Baghouses are the most effective of the three, but they do have some operational limitations.

There are two major types of gas-cleaning equipment used to remove acid gases from the combustion gas stream: wet scrubbers and dry scrubbers. The Clean Air Act amendments of 1990 require a 50-percent reduction in sulfur dioxide emissions—an acid rain precursor—by the year 2000, so these devices have gained in importance.

Wet scrubbers absorb chemical pollutants from the flue-gas stream using liquids. Particulates may also be reduced during acid-gas absorption, but not with the efficiencies offered by baghouses or precipitators. Corrosion-resistant materials are often required in the scrubber and associated equipment because alkaline reagents are commonly incorporated in wet scrubber design. There are also appearance problems at times because evaporation can cause thick white "smoke" to be emitted from stacks.

Dry scrubbers are currently the most effective form of acid-gas reduction equipment. They atomize caustic slurries that then become neutral calcium or sodium salts when they interact with the waste-gas stream. Particulate control devices then remove these materials when they collect the dried salt droplets.

Since much less water is used in this process, no visible emissions flow from the stack after scrubbing.

Recycling and Composting

Recycling has long been practiced in the United States, and, during World War II, collection of aluminum and other metals for use in the war effort was highly successful. Composting also has long been practiced privately for backyard gardens and farms and in some rural agricultural areas. In the 1980s, in attempts to find the "gold in garbage," advanced technologies were applied to solid waste streams to remove metals, glass, and other recyclable materials before the remainder was incinerated for energy production. In the 1990s, recycling will move much more heavily to households, which produce half to two-thirds of municipal solid waste in a typical region, according to EPA estimates. Curbside recycling programs have been increasing dramatically throughout the United States as communities attempt to deal with the loss of landfill capacity. Sorting items such as different colors of glass, cans made of different metals, newsprint and coated papers, cardboard cartons, and plastics is required so that collected materials can be processed for reuse. Many communities require sorting for curbside pickup or, if there is a central drop-off point, provide separate labeled containers. Materials recovery facilities (MRFs) can also provide central sorting. Roughly half of the MRFs in the United States are highly mechanized, while the remainder depend on labor-intensive hand-sorting processes.

In the past, composting was little used in the United States and Europe (except for sewage sludge) because other modes of waste management were less expensive, and cheap artificial fertilizers were readily available. It is growing rapidly, however, as landfills reach capacity or close down and public opposition discourages further expansion of incineration. Composting can also be a convenient way for communities to handle peak loads, especially with the addition of yard and lawn wastes during summer and fall months. There is a potential outlet for compost as nourishment for potted plants within the $4.7 billion nursery and greenhouse market. Field-grown shrubs and trees could also make use of compost to replenish soil humus removed during harvesting and provide organics for producing good-quality soils.

Any biodegradable material can be composted, including food and yard waste, paper, and wood, which make up more than 65 percent of the solid waste stream, according to Franklin Associates. Before composting, these must be separated from materials that cannot be composted such as metals, glass, rubber, leather, textiles, and plastics. If properly processed according to EPA requirements, composting produces a disease- and seed-free soil additive that enhances aggregation, improves porosity and aeration along with water infiltration and retention, and decreases crusting. It has little value as a fertilizer, however.

Composting requires four basic steps: preparation, digestion, curing, and finishing. Preparation requires removing any items that cannot be composted, inorganic components, as well as any valuable materials, and then shredding the remainder. Digestion occurs in open rows (called windrows) or in closed containers. In either case the material is ground; moistened; and reoxygenated periodically to ensure aerobic conditions, remove heat generated during decomposition, and prevent foul odors. Curing takes from two weeks to several months depending on operating procedures and local conditions. After digestion is complete, it is necessary to break down any remaining lignin and cellulose. Finishing is essentially preparing the compost for its intended use by drying, grinding, or some other process.

Waste Minimization

There are two basic goals in minimization: to reduce the quantity of all waste, and to reduce the amount of toxic materials in the waste stream. Although this is the preferable method of dealing with solid waste reduction, efforts to apply it have been minimal. One drawback is the need for public participation to develop successful programs, and this is difficult to achieve without some sort of incentives. Also, in the 1980s, government efforts were directed toward other issues, such as hazardous waste. There are three categories of waste that must be considered: home scrap, which is both produced and reused at the original manufacturing facility; prompt industrial scrap, which accumulates at intermediate processing facilities in the production chain and can be returned to the primary production facility for reuse; and old scrap or postconsumer scrap, which is discarded by the ultimate consumer of the product.

Waste minimization depends primarily on altering the attitudes and behavior of consumers, distributors, and producers of goods rather than advances in technology. Attempts to modify behavior throughout the marketing chain must take account of the costs involved in any options that are considered, not just to each participant in the process but also to governments from the local to the federal level. Potential cost savings can be a powerful incentive, while higher costs can trigger strong, organized resistance. Education also plays a major role, particularly by encouraging consumers to change their lifestyles. To make sales, vendors must respond to consumer demands, so that, as the environmental movement gains momentum, more and more changes are occurring. Corporations such as McDonalds have responded to environmental criticisms by changing packaging materials, making cartons lighter and thinner, changing some packages from polyurethane to recyclable cardboard, and encouraging recycling of some items by offering separate disposal bins. Large appliance and consumer electronics companies such as Whirlpool and General Electric are beginning to design products for easier disassembly to aid recycling. Reduction of composite materials, adhesives, and one-way fasteners helps make these items simpler to take apart and sort.

The greatest opportunities for reducing solid waste lie in packaging, and many changes are already occurring. Coffee packers have replaced tin cans with a brick package technique that is 85 percent lighter. Procter & Gamble has taken a leading role in repackaging consumer goods to make them less harmful to the environment. Crisco Oil bottles now use 28 percent less plastic for the same quantities, packaging has been reduced 80 percent for disposable diapers, and recyclable plastics are now used for detergent containers.

Incentives and disincentives at each level of the market, from producer to final consumer, as well as outright bans, are also options. Economic incentives or penalties tend to be small for each consumer, even though the cumulative effect of their decisions might be huge on the society as a whole. Therefore mechanisms that have greater impact on producers or distributors may prove more effective. The German experiment that began in 1991—forcing retailers to accept discarded packaging from consumers right at the store—is expected to lead to greater dependence on reusable or recyclable packaging. Retailers are likely to put pressure on vendors to reduce packaging and to find more ways to distribute products in bulk.

TECHNOLOGY TACKLES ENERGY PROBLEMS

One area where technology can play a major role is in increasing energy efficiency, particularly in the United States, which is the world's largest energy producer. Greater efficiency in both generating and using energy could have numerous beneficial environmental effects. It would help to conserve the world's limited supplies of fossil fuels while greatly reducing the production of greenhouse gases and acid-rain precursors. More efficient industrial processes would save money for producers and help make U.S. businesses more competitive in world markets. Studies have shown that U.S. industry is only about half as energy-efficient as competitors in Germany and Japan, for example.

The major emphasis on energy efficiency in the United States has been on conservation, the demand side of the electric power business. More energy-efficient buildings and building components, such as windows, appliances and heating systems, are gaining acceptance in new construction, for example. But getting industry and consumers to choose more energy-efficient machinery and appliances is proving to be a difficult challenge. Governments have sponsored programs urging more energy-efficient choices, conducted mainly by electric utilities. The power companies encourage their customers to conserve and thus help forestall the need for added capacity. Even if successful, such programs would take decades to replace all the equipment and devices driven by electric motors. Meanwhile, we can achieve more effective energy efficiency and less pollution by shifting the focus toward the supply side of electrical generation by installing more effective technology.

Some 30 percent of the world's energy goes into producing electric power. The proportion is an even higher 35 percent for North America and a little below that for the United States. About two-thirds of this energy comes from fossil fuels, 20 percent from hydroelectric projects, and most of the remainder from nuclear power. After the oil shocks of the early 1970s many U.S. utilities switched to coal, and, because the United States has huge supplies of domestic coal, it makes up some 56–57 percent of the fossil fuel used for producing power. Many utilities have installed systems suitable for a variety of fuels so that they can better cope with either shortages or excessive price fluctuations. The concern over greenhouse gases and their possible role in climate change is likely to result in an increased role for electricity as a primary source of energy. Although there is considerable publicity about electric cars in the consumer sector, the shift will be much greater for industrial processes, including more electric minimills for producing steel from scrap, infrared heating for drying, and robotics and automation for manufacturing.

To meet this growing industrial demand, electric utilities need more fuel-efficient, less-polluting generating systems. Not only would this greatly improve the utilities' environmental performance; more efficient generation would also be economically rewarding in the long run. A number of excellent high-efficiency, lower-emissions technologies for electric power generation have emerged from R&D, but utilities in the United States have been slow to embrace them. A number of factors—including the regulatory climate, the recent economic downturn, and cautious management within the U.S. power industry—have all served to slow improvements here while European and Japanese manufacturers take over the leadership in advanced generating technology.

Although important new generating technologies are in the advanced research phase, there are cleaner, more efficient systems available now to replace traditional steam boilers. Chief among these is the gas turbine combined-cycle (GTCC) system, which is cost-competitive and has already been installed and operated reliably in dozens of power plants in the United States and Japan. Retrofitting existing plants with GTCC systems can reduce carbon dioxide emissions by some 25 percent. These systems burn oil or natural gas in a combustion turbine with any waste heat used to drive a steam turbine. Since both cycles produce electricity, the overall fuel use and total emissions drop for generating the same quantity of electric power. Converting fuels from coal or oil to medium-carbon natural gas has the potential to cut emissions another 25 percent. Improved operating efficiencies for such plants can almost pay for the capital costs for installing them.

In countries where coal is abundant such as the United States and China, another transitional technology that is somewhat more costly is the integrated gasification combined-cycle (IGCC) system. Coal is heated in pure oxygen to generate gases—mainly hydrogen and carbon monoxide—that can have sulfur and particulates removed before being burned in combustion and steam cycles like those in GTCC systems. Because of the use of coal, IGCC systems are not quite as good at lowering emissions.

Both of these technologies offer significant advances in the efficiencies of converting fuel energy into electricity. Whereas steam-boiler and steam-turbine plants typically operate at 20- to 38-percent efficiency, GTCC plants built over the next 10 years are expected to perform at 46-percent efficiency and IGCC plants at 39-percent efficiency. By the turn of the century or so, there may be even more efficient technology available using what are called "humid" turbines. A mixture of hot air and steam burn natural gas or synthetic gas (syngas) made from gasified coal in a single turbine, reaching efficiencies of above 50 percent. Since only one turbine will be needed, capital costs should also be lower.

Several technologies are being explored for cleaning up coal for energy production, including fluidized-bed combustors with high-temperature desulfurization and particulate removal with ceramic filters. Gas must be produced that, when combusted, has very low sodium and potassium content and is acceptable for use in a gas turbine. Designers face severe materials problems in building suitable high-temperature coal-gasification systems. In a pressurized fluidized-bed system, sulfur is removed in the bed within a temperature window of 450 to 650 °C, while particulate removal takes place at 850 °C. Work is now progressing under a DOE project called Coal 2000 on a different type of cycle that puts coal gases through a heat exchanger to produce a very clean, high-temperature air stream for running the turbine. This approach avoids the problems associated with cleaning up coal combustion products, but instead depends on the development of suitable high-temperature heat exchangers.

An alternate technology that depends on electrochemical reactions rather than combustion for producing electricity is the molten carbonate fuel cell. Again fuels such as natural gas or gasified coal could be used in a fuel cell, which oxidizes hydrogen and carbon monoxide in an electrochemical reaction to produce electric power. Pilot projects have demonstrated the viability of this technology.

While fuel efficiency of electric generators is important, another 10 percent of the electricity generated by utilities is lost in transmission. To help cut these losses, there is increased long-distance transmission using direct current (DC) because resistance is lower for the same size and capacity of conductors than for alternating current (AC). Long-distance DC has become a more viable option because costs have been declining for the conversion equipment needed to switch between DC and AC for transmission to customers. Research also continues on superconductors capable of serving as transmission lines, but building practical superconducting transmission systems is likely to be a few decades off. Superconductors offer essentially zero resistance, thus promising much lower transmission losses. To make them practical, however, researchers need to find materials that can easily be formed into conductors capable of carrying high currents without losing their superconductivity. Cost-effective methods also will be required for cooling the conductors. Even if suitable materials are found that turn superconducting at higher temperatures, considerable cooling still would be required.

Another potential source of energy is the "waste heat" produced at central power stations and large industrial plants. Cogeneration is a process of using the exhaust steam from electric generation to supply energy for other uses. Dow Chemical, for example, has reduced energy consumption by 50 percent over the past decade largely through building its own central electric generating plants and using steam from these plants to supply energy to other operations. In Germany, any new power plants must include plans for using exhaust steam either for industrial processes or district heating. Waste heat could also be used to run cooling equipment, in some cases displacing refrigerants consisting of ozone-layer-damaging CFCs.

Because the greatest growth of electric generating capacity in the next few decades is likely to be in the developing world, international efforts will be important to ensure that these more fuel-efficient, cleaner technologies are chosen over more wasteful and polluting methods. This will be particularly important in coal-rich areas such as China. Hydroelectric power also offers an option that avoids emissions problems, but again there is great potential for environmental abuse unless the industrialized nations find ways to offer the expertise and assistance needed to ensure that such projects do not destroy resources and damage ecosystems. China's plan for the Three Gorges Dam on the Yangtze River, for example, would displace a million people to create a 370-mile-long reservoir in an area that is one of the nation's chief tourist attractions. This project may be stymied by costs, estimated at tens of billions of dollars, as well as by opposition from environmentalists and large numbers of peasants who would be uprooted. The Chinese government would have to borrow heavily to finance such a major project, and managers of the major lending institutions, such as the World Bank, have recently begun to steer clear of projects with such high social and environmental costs. It is possible that, if international rules encourage tradable permits to help reduce emissions of carbon dioxide and other gases into the atmosphere in the future, this might help developing nations to finance less costly hydroelectric projects.

ENERGY ALTERNATIVES TO FOSSIL FUELS

Nuclear power also offers an energy source without the emissions problems of fossil-fuel plants, but issues such as safety, reliability, waste management, and protection of nuclear materials from diversion for military use must be addressed by the international community. Considerable effort will be required just to remedy the legacy of past errors in the nuclear arena, including unsafe and poorly designed plants in the former Soviet bloc, mishandling of nuclear waste, and contamination cleanup necessitated by poor regulatory oversight of military facilities in the United States. Industrialized nations have increased their demand for compliance with these issues and need added assurances that they will be properly addressed if developing nations choose to increase their use of the nuclear option.

Technology is also advancing steadily for an array of "soft energy" options that are likely to become much more important in the next century. Wind power at 4 major sites in California generated about 1,500 megawatts (MW) over a 10-year period at costs in the range of 7 to 9 cents per kilowatt-hour (kWh), compared to a cost of about 5 cents per kWh for coal-based generation. If the costs of cutting down carbon dioxide emissions for coal plants were included, however, wind power would be economically competitive. Multiple new sites for an expansion of wind generators, in areas such as South Dakota, New Mexico, and Montana, are now under consideration. Although Europe only has about 250 MW of installed wind machines, the governments and utilities of several European nations—Denmark, Germany, the Netherlands, the United Kingdom, Italy, and Greece—have plans to install wind farms that would result in 3,000 to 4,000 MW of wind power in the EC by the turn of the century. Wind generators also have potential either for connection to power grids or for remote villages, where backup could be provided by diesel generators and battery storage.

Photovoltaic (PV) systems, in which current is generated by panels of semiconductor solar cells, are another promising technology. There are about 25 MW of grid-connected PV systems in the United States, with considerably less in Europe. High capital costs bring the price of PV-generated electricity to about 30 cents per kWh, which has limited their use. These costs have come down steadily, and continuing research is projected to bring them down to the range of 15 cents per kWh, which would make them more competitive. If a $250 per ton carbon tax were imposed, for example, this would put a penalty of about 10 cents per kWh on a 35-percent efficient coal-burning power plant. PV systems can be highly modular and do have advantages for small, isolated applications, such as in weather stations or, again, in remote villages.

Another promising source of future energy beyond fossil fuels is the heat within the earth, which might be tapped by hot dry rock (HDR) or magma geothermal energy systems. Both of these approaches employ closed-loop heat mining concepts to extract heat from accessible portions of the earth's crust. They would emit no effluents while providing uninterrupted power that would not be dependent on the weather. Designs for such systems are inherently simple, and they should be safe to operate, without danger of explosions, catastrophic failure, or leakage of harmful contaminants.

The concept being pursued for HDR is to drill holes 3–7 kilometers (km) (10,000 to 25,000 feet) deep in impermeable rock where temperatures range from 200 to 300 degrees C. Then the rock would be fractured deep in the hole using hydraulic methods. This would create artificial permeability within a underground reservoir. Water then could be pumped down to absorb heat and brought back to the surface where the heat would be extracted in a closed-loop system. DOE is funding work on HDR by Los Alamos National Laboratory while a similar agency in the United Kingdom sponsors work at Camborne School of Mines in England. The DOE is also sponsoring work on magma systems by Sandia National Laboratories. Potential HDR resources are huge and are well distributed all over the world. Estimates are that high grade HDR resources (geothermal gradients higher than 50 degrees C/km) could be tapped at break-even prices of 5–6 cents per kWh, but improved technology would have to be developed to make lower-grade resources commercially competitive at current prices for electric energy.

The technology required for high- and mid-grade (gradients of 30–60 degrees C/km) HDR resources could be provided by straightforward extensions of existing drilling and reservoir-stimulation techniques. A large-scale effort might allow such systems to be implemented in 10–15 years.

BATTLING THE BARRIERS TO INNOVATION

Because environmental technology is developed under a regulation-driven system at present, there needs to be cooperation between the public and private sectors to encourage innovation and the necessary investment. In the United States, the relationship between government and business is frequently more adversarial than collaborative. There are strong feelings that government should not "interfere" in business, and that, where government does intrude, the results are likely to be damaging rather than helpful. These underlying attitudes sometimes lead to difficult, even stormy relationships between regulators and the regulated. Where powerful interests are threatened by proposed legislation or regulations that would curb their activities or make them more costly, lobbying and carefully directed campaign contributions are often invoked to shape the outcome. The interests of the general public sometimes become submerged in this process. In the early days of environmental legislation, most of this pressure came from industry, but in recent times powerful environmental groups may have also exerted strong influence. Recent maneuvering in the Pacific Northwest, which became characterized as a battle between the spotted owl and jobs in the lumber industry, illustrates the emotion-laden distortions that can result from such a clash of interests. The environmental industry must operate within this culture, which may divert attention from substantive issues, often creating barriers to innovation rather than encouraging it.

In 1980, the Reagan administration focused its efforts on reducing the size and scope of the federal government and, in particular, its influence over business and industry. This translated into a laissez-faire policy, reducing government intervention in the marketplace. Industrial policy—the idea of using government power and taxpayer money to stimulate key industries through support of commercial research—became anathema. The Bush administration continued this hands-off approach to the marketplace. In his presidential campaign, Bush derided industrial policy as "picking winners and losers."

Recently, however, due to growing foreign competition to U.S. industry and increasing foreign government support of that industry, industrial policy is gaining acceptance. Paul Tsongas, a 1992 presidential candidate, charged that the Republican mania for free markets is dangerously out of date. Today, foreign governments keenly nourish their own private industries. "American companies," said Tsongas, "need the U.S. government as a full partner."

Bush administration members split on the issue of industrial policy and encouraging innovation. Some argued that the United States must find some way to spur technology development if American business is to compete with their foreign counterparts. Others feared that government support of commercial research would lead to federal interference in the free-market system, and they preferred a macroeconomic policy approach, such as cutting the tax on capital gains, which, hopefully, would encourage more investment in innovative enterprises. In a March 1990 speech, Bush softened his views by pledging to work with industry to promote "precompetitive" work on "generic technologies that support our economic competitiveness and our national security."

Translated, "generic" refers to technologies that could have broad applications across a number of industries. "Precompetitive" means that the federal government would not be part of a program to develop a specific commercial product or manufacturing process for a particular industry or company.

The term "national security" is also one that has recently entered into this debate. In the past, this term has been viewed in purely military terms. However, national security has increasingly been used to describe economic security and the promotion of advanced commercial technologies necessary for maintaining the competitive status of U.S. industry. This new view appears to have been much more warmly embraced by the Clinton administration.

The federal government has many tools at its disposal for promoting innovative technological development. The most immediate way in which government can influence the market is the **tax incentive.** Tax incentives can speed the adoption rate of new technologies by lowering the overall implementation cost to the user and, therefore, lowering the financial risk. The government can offer incentives directly to the innovator of the technology. For example, since 1981 the federal government has allowed industry a 20-percent R&D tax credit. The government can also offer tax incentives directly to the end user of the new technology. For example, tax credits could be offered to hazardous waste site owners on the funds spent to remediate their sites. This would help to increase the rate at which sites are cleaned up in this country without biasing the market towards any specific technologies. However, if the government wanted to focus efforts on specific types of technologies, it could offer greater tax credits for monies spent on certain technologies—for example, it could promote recycling, utilization of clean energy sources, complete detoxification of remediated wastes, or beneficial reuse of the waste. This approach is contrary to the Bush administration's policy of not picking winners and losers and could result in unfair bias toward specific existing technologies. Future innovations would be faced with even greater barriers to entry than already exist. This would be best avoided.

Other, less used, methods of support include: the use of national laboratories for private research; funding of industrial research either directly or through universities; and encouraging consortia which bring industry, government, and university researchers together to explore specific scientific research.

Researchers at national laboratories can become involved in commercial technology development in primarily two ways. Under the Stevenson-Wydler Technology Innovation Act of 1980, federal laboratories are encouraged to find commercial applications for their work. Previous to that act, a great deal of useful technology was being developed for the government, but it was just sitting on the shelf. Under the Federal Technology Transfer Act of 1986, industry and government researchers are allowed to collaborate on technology development. Through these two acts, the work of some of the 700 federal laboratories could be directed toward developing new technologies. One federal laboratory, the EPA's research facility in Cincinnati, Ohio, is already involved in remediation technology development through the SITE program.

Research centers created specifically to combine government, industry, and university resources have also proliferated over the years. Around the country, there are 18 federally funded engineering research centers, 11 science and technology centers, 3 "Hollings centers"—named after Ernest Hollings, senator of South Carolina—devoted to technology transfer in manufacturing and various other government-sponsored ventures. More are on the way, with a target of 12 Hollings centers, for example, to be placed strategically in different regions of the country. The Clinton Administration now has plans to extend this to 150 centers. The 29 engineering and science centers together cost the federal government $60.6 million in fiscal 1989, of which $7.5 million went to the Hollings centers.

The federal government can fund industrial research either directly or through universities. Direct funding for hazardous waste remediation technologies is unlikely since this method of support is used almost exclusively for military contractors. Funding, however, of industrial research at universities is a viable option. A 1982 National Science Foundation (NSF) survey found that industry benefited from university research relations through: access to students and professors; access to technology for problem-solving or obtaining state-of-the-art information; prestige; economical use of resources; support of technical excellence; and proximity and access to university facilities. University researchers benefited from their relationship with industry research in a number of ways: access to scientific and technical areas where industry has special expertise; the opportunity to expose students to practical problems; the use of earmarked government funds; and potential employment for graduates. Some universities, such as MIT and the University of Utah, now have programs that encourage researchers to commercialize potential products coming out of their work, sometimes with the university gaining a minority ownership position to compensate it for providing the facilities and support needed to develop the technology.

Some people criticize the use of academic institutions for industrial research. A report by the Organization for Economic Cooperation and Development (OECD) states that industrial research at the university level is more practical, short-term, and commercial in its orientation. It warns that the industrialization of research threatens the accessibility of research results. As a result, this "compromises the primary function of academic institutions...to serve the broad public interest."

Finally, government can promote development of innovative remediation technologies through encouraging **consortia**. These cooperative ventures can include industry, government, and academic researchers. The benefits include a significant reduction in the time and cost of capital-intensive and multidisciplinary research. Since the passage of the 1984 National Cooperative Research Act (NCRA), which encourages collaboration by shielding such ventures from antitrust litigation, more than 173 consortia have registered with the U.S. Department of Commerce.

Overseas, government-sponsored, and often government-orchestrated, consortia have proven successful. Some, such as the Esprit and Eureka programs in Europe and Japan's Erato, Optoelectronics, and Monbusho programs, have been aimed at fostering the development of generic technologies that could then be commercialized by industrial participants. In Japan, the Ministry of International Trade and Industry coordinates programs such as these, with the goal of raising the overall competitiveness of Japanese industry in world markets. In Europe, the airbus program is a highly subsidized program involving manufacturers in various nations to create a world-class trans-European company to build and market commercial airliners. These programs are all examples of highly successful efforts to develop key domestic industry sectors that could compete strongly against foreign competitors.

The level of U.S. government involvement in cooperative agreements has ranged from minor steps, such as easing antitrust restrictions, to major contributions, including direct financial assistance. Federal involvement in the semiconductor consortium, Sematech, amounted to contributing half of the entire $200 million price tag. This funding came from the Defense Advanced Research Projects Agency (DARPA), which feared that weak U.S. industries threatened national security. DARPA has also provided $30 million over three years for high-definition television (HDTV) research. The Commerce Department has also initiated its Advanced Technology Program, which will distribute $10 million to fund future industry-led consortia.

Critics contend that consortia have, as yet, been ineffective in promoting technological competitiveness. They feel that declining competitiveness of many U.S. industries is not due to deficiencies in basic research but to an inefficiency in transferring technology to the marketplace. Videocassette recorders, semiconductors, and televisions are all examples of markets in which U.S. scientists pioneered technological breakthroughs only to have foreign competitors dominate the market through better and faster entry. This inefficiency will only be exacerbated by the formation of a consortium since competitors cannot really cooperate at a level of research so close to the marketing of the final product.

The hazardous waste business illustrates many of these problems. With long time scales for payoffs, public investment has essentially dried up for this sector. The EPA may take ten years to issue a license for the cleanup of a particular site, and then it dictates which technology should be used. In approving technology proposals, the EPA is concerned about making certain that any hazardous materials have been eliminated. Demonstrating compliance can prove difficult for some innovative technologies, such as bioremediation. Those paying for the cleanup must balance limited budgets for up-front demonstrations against hopes for long-term savings through technological advances.

Although the EPA has been working to develop new guidelines that would help overcome some of the barriers to innovative technology, the experience of those in the business continues to be trying at times. In one case, Halliburton NUS developed a remediation technique for basic extractive sludge treatment, called BEST. It cost-effectively dewaters and removes organics from contaminated soils and sludges by using the reverse solubility characteristics of triethylamine. The technique was used in an emergency situation in Savannah, Georgia, in 1988 to process several tens of thousands of yards of contaminated sludge. The company developed at its own expense a transportable treatment unit that could be used for pilot projects at contaminated sites to evaluate the cost-effectiveness and efficacy of the treatment for soil or sludge remediation. It applied for and received a permit from the EPA for operation of this pilot unit. But when the company tried to use the unit for a 6–8-week demonstration in New York under its EPA permit, regulators there concluded that the company would have to apply for a special permit for that state. Managers did not know exactly what was needed for approval so they just began to submit data, waiting for the agency to issue an approval. The process took 8 months, an investment of $250,000, and 9 inches of submittals that had to be sent to 25–30 interested state agencies. Even after all that, an authorizing letter, rather than a permit, was issued. And this was for a 6–8-week pilot demonstration rather than an actual cleanup.

A comment of Peter D. Arrowsmith, president and chief operating officer of Halliburton NUS Environmental Corp., sums up one major structural problem that must be addressed in order to improve the pace of innovation in the environmental area: "When senior officers in our industry—and I'm one of them, unfortunately—spend more time with their lawyers than with their engineers or in reviewing work at a job site, we're clearly working on the wrong problems."

In conclusion, because traditional environmental policies fail to encourage innovation to protect the environment, alternate methods of stimulating new technology are required. Command-and-control regulation has not proven very effective at encouraging innovation, partly because of the high risk of markets being wiped out by regulatory shifts. Policy makers have tried to develop more market-oriented approaches that do not rigidly prescribe technology. They would like to reward those who find more cost-effective methods for reducing or eliminating pollution and penalize those who are less diligent with higher costs for the waste they produce. There are a wide range of environmental technologies being developed, but under the current regulatory system there is some difficulty in getting them tested, accepted, and out into the marketplace. The next chapter will look more closely at trends in individual environmental markets, in the United States and elsewhere, and will discuss opportunities for construction firms to participate in this rapidly growing marketplace.

CHAPTER 5

ENVIRONMENTAL MARKETS

Success in entering any market depends on understanding its dynamics as well as careful analysis of its direction and growth potential. Construction firms hoping to enter one or more sectors of the environmental marketplace, however, need to go beyond traditional market analysis and projections because of the special characteristics of this sector. Since every construction firm will be increasingly subject to environmental pressures in future projects, it is important for decision makers to be aware of these trends even if the company does not plan actual entry into the marketplace. They will need to understand the unique character of this business, gaining perspective on how market opportunities flow from public demands for environmental controls but then are shaped by changing directions in public policy making. To help in this learning process, this chapter looks at the fundamental differences between the commercial and environmental marketplaces. It includes discussion of the behavior of environmental markets in both the public and private sectors, describes the various segments of this rapidly growing marketplace, and develops business trends and forecasts for each segment. Since the construction business is becoming increasingly globalized, it will also cover trends and forecasts for major environmental markets around the world.

HOW MARKET DEMAND IS CREATED

Public pressure for a better environment has resulted in an expanding array of markets for environmental equipment and services. As noted in the previous chapters, the environmental sector differs from other types of businesses in that markets and their associated technologies are created and shaped primarily by regulation rather than by consumer demand. In a free-enterprise business climate, executives attempting to minimize costs resist spending on what they consider to be non-value-adding items. Under the conventional view, environmental protection such as pollution controls or waste-management systems are considered to be overhead costs, external to the main line of business, so most companies will avoid investing in this area unless they are forced to comply with specific regulations. However, when legislation is passed that defines such regulations, and enforcement provisions threaten violators with fines or even prison terms, affected industries begin to seek solutions to their environmental problems. This quest sets the stage for the emergence of a competitive marketplace to provide the controls or treatment facilities mandated by the law.

Contractors and equipment vendors with capabilities and expertise in the areas of concern will begin to offer a variety of products and services to meet regulatory requirements. As the market expands, contenders can use a number of factors to differentiate their offerings from competitors, including advantages in technology; depth and breadth of experience of their organization and personnel; knowledge of applicable regulations; ability to meet tough timetables; and, often the most critical factor, lower costs.

In the public sector, buyers are typically federal, state, or municipal agencies. The public marketplace includes major remediation projects, such as the cleanup of Superfund sites or of waste dumps at military or other government installations, as well as installation of new facilities, such as sewage treatment plants, water treatment facilities, or modern landfills. In the early phases of most public projects, specifications are first developed carefully to define any work that must be done and the results that must be achieved. Then closed bids are solicited. The work is then most commonly awarded to the lowest bidder that can also meet all pertinent requirements, such as obtaining adequate bonding and liability insurance. In certain cases, the public agency may choose a higher bidder because of special expertise or technology, but such awards are often challenged by lawsuits, so clear-cut reasons are needed before most agencies will take such a step. After gaining experience with particular contractors who meet schedules and do excellent work, however, public agencies may sometimes decide on a solicited bidding procedure open to only a couple selected organizations, or the agency may even negotiate fees with a single organization for a new project. Environmental jobs may be contracted to firms that do none of the actual work. The prime contractor may instead act as project manager, employing an array of specialized subcontractors to perform individual tasks, such as transporting hazardous waste or treating soil and groundwater. Some public agencies, such as the U.S. Army Corps of Engineers, are quite sophisticated in the environmental marketplace and may act as their own project managers, breaking up large projects into specialized subtasks, and then inviting firms specializing in those areas to bid only on that portion of the work, such as transporting hazardous waste or installing particular types of treatment plants.

In the private sector the environmental business may operate quite differently. Rather than soliciting bids, a private contractor may employ a set of evaluation factors, such as capabilites, experience, and financial stability, before choosing an equipment supplier or an organization to do required work. In the case of an industrial company that needs to ensure that its plants will not be cited for environmental violations, selecting a contractor may depend heavily on confidence that the chosen organization can ensure compliance to applicable regulations and reduce any risks of long-term liabilities. These factors may be even more important than the specific technologies offered or even costs. Fees might be negotiated after a contractor has been selected, based on a detailed assessment of the work to be done. In principle, the basic contract formats, in both the public and private sector, are the same in environmental projects as in any others. However, the nature and extent of risk, liability, and unanticipated complications make environmental contracts rather unique.

DYNAMICS OF THE ENVIRONMENTAL MARKETPLACE

There is a complex interplay of diverse forces across this unique marketplace, often far removed from the usual business considerations, starting with public demands for environmental improvements. Legislators and other policy makers respond to the demand for a better environment by developing the regulatory framework under which improvements will be pursued. Politics can sometimes become an overriding consideration. As mentioned earlier, congressional committee members from states that produce low-sulfur coal might favor legislation requiring emissions to be lowered by the use of the coal from their districts over some alternate legislation mandating emission reduction by cleaning up flue gases. Once laws are passed, regulators must interpret the legislation, developing policies into detailed requirements, seeing that these are implemented, and carrying out appropriate enforcement. In some cases, the regulations may get down to the level of individual polluters, requiring them to cut back or eliminate emissions or effluents within specified time periods.

As industry seeks to comply with environmental rules, companies with the requisite skills and hardware capabilities see the business opportunity and begin to respond by offering products and services that can help those affected to meet mandated requirements in a cost-effective manner. In some cases, new developments in technology may encourage regulators to stiffen limits. In other instances, legislators may call for levels of cleanup that are not yet attainable, and regulators may have to set timetables and challenge the environmental industry to come up with improved methods and technology capable of meeting the more stringent criteria. Companies developing and applying methods and technology within the marketplace must be attuned to directions in policy and regulation, recognizing that such shifts can create dramatic changes in a business that already involves very high risk factors.

In spite of the drawbacks—high costs of entry and risks such as regulatory shifts and long-term liabilities—a highly diversified, rapidly expanding environmental industry is taking shape to meet the needs of a wide array of regulations. So far, it is not a very well-defined industry, but some broad, general categories can be identified before describing narrower, more well-defined market sectors. Some portions of the industry have a long tradition in areas such as wastewater and sewage treatment and even equipment for controlling air pollution. Filtration units for flue gases have been around for nearly 100 years, with the Cottrell precipitator appearing at about the turn of the century when the air pollution marketplace began to take shape. The rapid rise in legislation beginning in the 1970s, however, has certainly added tremendous impetus to the growth of the environmental market.

Expanding opportunities have also brought a steady stream of new entrants, most often in traditional product or system categories, but occasionally with improved technology or methods. The requirements for controlling pollution, treating effluents, disposing of various types of waste, and even digging up buried harmful waste and rendering it harmless are extremely diverse, so that companies from a wide range of industries are entering the marketplace. For example, one area of emerging technology is bioremediation in which selected or even genetically engineered organisms may turn hazardous or undesirable waste into harmless substances; thus, a new breed of firms with a strong footing in the biotechnology field is vying for entry into this potentially lucrative hazardous waste remediation market.

Construction companies—of all sizes and types—have also entered or expanded their activities within the hazardous waste remediation marketplace, from site planners and engineering services firms to contractors looking for additional work for earth-moving equipment. The trend has been gaining pace as conventional construction work has slowed down during the economic recession that has extended into the 1990s. One major attraction is a profit level as high (on the average) as for any industry in the United States, except perhaps pharmaceuticals. But there are important drawbacks that must be considered before entering this marketplace. Many companies want to get into a market when it's growing, stay with it as long as business remains healthy, and then get out when things begin to tail off. Under present liability laws, however, any organization that generates, manages, transports, or deals with hazardous materials in any way is potentially liable forever. That means that, once a firm enters any part of the business where hazards exist, there may be significant legal barriers to exit. Thus, entry into the environmental marketplace should be a carefully considered, strategic decision.

STRUCTURE OF THE ENVIRONMENTAL MARKETPLACE

We can view the environmental marketplace as a continuum, although the demands within each sector are usually so distinct that there is very little overlap in the list of competitors across sectors. At one end are production technologies and processes that eliminate and/or reduce production of environmentally undesirable by-products. This is the source-control approach. Then there are controls designed to prevent pollution—end-of-pipe solutions—that require treatment facilities for waste that is created. Finally, there are methods for dealing with waste that has been improperly disposed of in the past. Following is a brief description of major categories, or market sectors, within the total marketplace.

The **remediation** sector deals with corrective action, cleaning up previously deposited waste and treating contaminated soil and groundwater. This sector also includes asbestos-removal projects. The **pollution control** sector deals with equipment and methods employed to treat various types of effluents emitted by smokestacks or drainage pipes. The **waste-management** sector includes systems for sorting solid waste for resource recovery, so that selected items may be recycled or composted. Waste that cannot be reused may be reduced through waste-to-energy incineration or buried in sanitary landfills. A relatively new sector is **hazardous waste clean-up and treatment.** This includes both specialized remediation to clean up existing sites and pollution control activities to prevent or treat newly generated hazardous substances.

These market sectors have developed within the current regulatory-driven environmental framework. But, as noted earlier, an emerging sector that will become much more important in the future is dedicated to **source control**—avoiding the creation of waste or pollution in the first place. This constitutes a widely distributed sector, including add-on equipment, for such tasks as cleaning up industrial solvents, as well as treatment facilities allowing by-products to either be reused within a process or else marketed for other purposes. Examples of such by-products finding external markets are the sulfur removed during the treatment of flue gases or sewage sludge, which may be sold as a fertilizer after it has been processed to remove heavy metals and other toxic components. An important factor in generating growth in this sector is wider acceptance within top management ranks that waste constitutes a significant opportunity not just for cost savings but also as a source of new marketable products.

Within each of these sectors, diverse organizations do the necessary consulting and other analytic services, such as planning, assessment, and feasibility studies, plus testing and design. Then there are specialty firms that perform the actual fabrication, construction, and installation. Some companies only manufacture equipment, subsystems, or even turnkey systems for the environmental market. In some cases, particularly with hazardous waste, transportation is also a vital function. Once systems or facilities are built or installed, there is also a need for organizations that can do operations and maintenance work.

The *Environmental Business Journal* has identified fourteen industry segments. Table 5.1 shows these segments, with corresponding numbers of public and private firms in each segment along with their estimated annual revenues.

Skills and advanced technology developed in these markets in the United States will also be applicable to future cleanup operations in other parts of the world, including eastern Europe and the former Soviet Union, as well as much of the developing world, where resources are lacking to deal adequately with mounting waste and pollution problems. Therefore, those firms that gain competency in various aspects of remediation will undoubtedly find future opportunities abroad, in some cases through direct contracting, but probably more commonly through partnerships and technology-sharing arrangements with foreign companies. However, it is important to recognize that, even though such markets appear to have huge potential, many of these nations do not have the funds

Table 5.1 U.S. Environmental Industry Segments—1990

Industry Segment	Public Companies			Private Companies			Total Revenue (Estimated) ($ Billion)
	Approx. Number of Companies	Total Annual Revenue ($ Million)	Average Rev./Co. ($ Million)	Approx. Number of Companies	Total Annual Revenue ($ Million)	Average Rev./Co. ($ Million)	
1. Analytical services	7	133	19	1,600	1,700	1.0	1.8
2. Solid waste management	15	11,096	740	4,200	17,500	4.2	28.6
3. Hazardous waste management	35	4,555	130	2,400	8,700	3.6	13.3
4. Asbestos abatement	14	1,313	94	3000	2,700	0.9	4.0
5. Water infrastructure	27	6,561	243	3,100	7,400	2.4	14.0
6. Water utilities	13	1,620	125	24,000	9,900	0.4	11.5
7. Environmental consulting/engineering	28	4,406	157	7,600	7,800	1.0	12.2
8. Resource recovery	21	4,566	217	5,100	12,600	2.5	17.2
9. Instrument manufacturing	12	659	55	500	1,100	2.2	1.8
10. Air pollution control	16	3,439	215	1,600	2,000	1.3	5.4
11. Waste-management equipment	17	1,728	102	5,000	7,500	1.5	9.2
12. Environmental energy sources	10	192	19	800	1,600	2.0	1.8
13. Diversified companies	5	1,990	398	2,000	5,000	2.5	7.0
14. Conglomerates*	17	3,509	206	500	525	1.1	4.0
Totals	237	45,767		61,400	86,025		131.8

*Only 3.5 percent of conglomerates revenue is included as "environmental revenue."
Source: *Environmental Business Journal*, April 1991.

to pay for needed facilities. They may also lack the infrastructure needed to support state-of-the-art technology, such as spare-parts suppliers and trained technicians. Yet they may resent being offered what they consider "old" technology. Either their economies will have to develop so that they can afford badly needed environmental work, or the industrialized world will have to set up mechanisms to finance them so as to avoid further global environmental degradation.

MAJOR GROWTH IN U.S. ENVIRONMENTAL MARKETS

Growth of environmental markets is expected to be huge in the United States alone, no matter what happens overseas. There are opportunities in each of these sectors for different portions of the construction industry. Dramatic expansion is expected over the next few years in the remediation portion of the marketplace, as hundreds of thousands of hazardous waste sites are cleaned up and as such problems as leaking underground tanks and groundwater contamination from old landfills must be addressed. By their nature, remediation markets will be temporary, although the tremendous magnitude of necessary cleanup tasks ensures that they will remain active well into the next century.

One estimate (by *Journal of Environmental Health,* July/August 1990) projects that the hazardous waste market will more than double in the next 5 years, growing from $11.5 billion in 1991 to $15 billion in 1992 and rising to $23.5 billion by 1996. DOE expects to spend $38 billion on cleanup projects between 1993 and 1997. By the year 2019, the Department's total expenditures for cleaning up contaminated waste sites and bringing their facilities into full environmental compliance may reach $150 billion, according to some estimates. Congressional overseers have recently challenged some of the very high figures for such cleanup operations, charging that a large portion of estimated costs are based on coverage for potential liabilities rather than the required work itself, and some adjustments in procedures may be needed to bring such costs down. DOD estimates that it will take approximately $14 billion for complete environmental restoration of its facilities.

The total environmental market was estimated to be $146 billion in 1991 (according to *Environmental Business Journal,* April 1991). Since the gross domestic product (GDP) of the United States was about $6 trillion in 1991, that makes the environmental market about 2.4 percent of total GDP. The EPA expects this percentage to grow further, reaching 2.7 percent of GNP. To put the size of this industry into perspective, the total value-added volume of the construction industry was estimated to be $143 billion in 1991, according to

Table 5.2 Projected U.S. Market Growth—April 1991

Industry Segment	Annual Growth (%)	Projected Revenue ($ Billion)					
		1991	1992	1993	1994	1995	1996
Analytical services	14	2.3	2.7	3.1	3.6	4.1	4.7
Solid waste management	8	30.9	33.4	36.0	38.9	42.0	45.4
Hazardous waste management	14	15.1	17.2	19.6	22.4	25.5	29.1
Asbestos abatement	4	4.2	4.3	4.5	4.7	4.9	5.1
Water infrastructure	10	15.4	16.9	18.6	20.4	22.5	24.7
Water utilities	4	12.0	12.5	13.0	13.5	14.0	14.6
Environmental consulting/engineering	16	14.2	16.4	19.1	22.1	25.6	29.7
Resource recovery	15	19.7	22.7	26.1	30.0	34.5	39.7
Instrument manufacturing	10	1.9	2.1	2.3	2.6	2.8	3.1
Air pollution control	16	6.3	7.1	7.9	8.9	9.9	11.1
Waste-management equipment	12	10.3	11.6	13.0	14.5	16.3	18.2
Environmental energy sources	8	1.9	2.2	2.4	2.7	3.0	3.4
Diversified companies	12	7.8	8.8	9.8	11.0	12.3	13.8
Conglomerates	8	4.0	5.0	5.0	5.0	6.0	6.0
Totals		146.0	163.0	180.0	200.0	223.0	249.0

Source: *Environmental Business Journal*, April 1991.

Engineering News Record, January 27, 1992 (when costs of materials and other resources are added in, total construction expenditures were about $430 billion).

Even though there is already a sizable environmental market, it appears that some industry segments will become substantially larger over the next few years because of double-digit growth rates. The *Environmental Business Journal* (see Table 5.2) estimates that the environmental consulting/engineering and air pollution control industry categories will see 16-percent growth rates, with resource recovery right behind them at 15 percent, and analytic services and hazardous waste management at 14 percent. These are in sharp contrast to a more traditional sector such as water utilities, projected to grow at only 4 percent a year.

OPPORTUNITIES IN ENVIRONMENTAL MARKET SECTORS

A review of the characteristics of each environmental industry segment, using somewhat broader categories than those in Table 5.2 (combining the drinking-water treatment and purification portion of water infrastructure with water utilities in a single water supply category, for example), will help to indicate some of the reasons for these growth projections. The categories we will cover are environmental assessment, hazardous waste, solid waste, sewer/wastewater, water supply, air pollution abatement, energy, and petrochemicals. The latter two categories, energy and petrochemicals, are not traditional segments such as hazardous waste or water supply, but environmental regulations play such a vital role in these industries that an analysis of them is critical to understanding the dynamics of the total environmental marketplace.

Opportunities for construction companies vary widely in different segments of the environmental marketplace. In some areas, such as wastewater treatment or sewage plants, construction firms have long been in the business and have built strong, entrenched positions. The market also has remained comparatively steady over the years. In others, such as remediation, there is a steady stream of new entrants, with many of them offering new approaches that may or may not prove effective enough to displace existing players. Rapid growth is the lure that has attracted so many contenders, many of them from the construction business, and even further expansion is expected as many Superfund sites move from planning to cleanup stage. The following is a brief overview of important trends in each major sector.

Environmental Assessment

Included in this fast-growth sector are both environmental engineering/consulting and testing. Confusion can arise if an attempt is made to isolate the amount of business done by companies within this sector. Just as is the case in the construction business, it is hard to separate firms that perform only engineering and consulting services from those doing environmental treatment and the construction and installation of treatment facilities. Many environmental engineering/consulting firms have added groups that do specialized treatment and other work, and a few environmental construction and treatment contractors have planning and consulting sections. Further confusing the breakdown is the addition of environmental services units to traditional broad-based engineering and construction firms such as Bechtel, Fluor Daniel, and Foster Wheeler.

There are good reasons why almost every firm wants to be in the fast-growth environmental services area. When analytic services are added to engineering/

consulting and testing, the total business for this sector is estimated to have reached close to $14.5 billion in 1991. Even though growth has slowed somewhat as the industry matures, from the burgeoning 25–30 percent annual growth rates of the past to a projected 16 percent, this still remains the fastest growing segment of the industry, according to *Environmental Business Journal*. Even at a growth rate of 16 percent annually, the business will double over the next 5 years. There are important attractions to this business even beyond this rapid growth. One is breadth of involvement: companies in this sector serve every segment of the environmental business spectrum. That means service organizations can be alerted to a broad spectrum of market opportunities. Early involvement is another advantage. Consultants and engineers are the first to be called in to do spadework such as conducting studies of hazardous waste sites or designing pollution-control systems. Being there first can often give a company an inside track to later business, and setting up specifications and the work to be performed can prove a tremendous advantage to a company that also wants to bid on implementing the plans it has devised.

The large traditional construction firms recently entering the environmental market expect to benefit as demand shifts from studies to implementation. This shift in demand is particularly noticeable in the hazardous waste sector, where increasing numbers of Superfund projects are now moving from investigation and feasibility studies to remedial action plans. "The market is increasingly moving in our direction," observes Roger Strelow, senior vice president of Bechtel Environmental Inc. Major engineering/construction firms such as Bechtel commonly deal with complex contracts take last several years to complete and reach into hundreds of millions of dollars or even billion-dollar levels. Many of the large government-financed cleanup projects that will soon be coming on-line seem ideal for these construction-industry giants.

Environmental testing, while only a small part of the total business sector at about $2.5 billion annually, is also expected to see growth rates of about 14 percent a year over the next 5 years.

Hazardous Waste

Over the past decade, the rapid emergence of the hazardous waste sector has had a strong influence in defining the environmental industry. Major legislation that shaped this market segment includes 1976's Resource Conservation and Recovery Act (RCRA), which established "cradle-to-grave" management of hazardous wastes from generation to final disposal, as well as 1980's Comprehensive Response, Compensation, and Liability Act (CERCLA, or the Superfund legislation), which directs the remediation of abandoned hazardous waste sites. RCRA deals primarily with active, regulated facilities while CER-

CLA applies to inactive or uncontrolled sites, although more recent amendments now allow RCRA to cover contamination cleanup much as CERCLA does.

While the hazardous waste market was only a modest $400–600 million in the early 1980s, it has grown at a annual rate of roughly 10 percent, far exceeding projections. The amount of work that will be needed to clean up contaminated sites seems staggering. The National Priority List (NPL) set up by the EPA under CERCLA includes 1,246 sites, and the OTA estimates that more than $500 billion may be required just to clean up these locations. Yet the GAO concludes that, under a more comprehensive listing, some 368,000 sites could qualify for cleanup. Added to these are growing numbers of state Superfund, DOE, DOD, and private-party sites, plus leaking underground storage tanks. Current estimates of the domestic hazardous waste market come in at about $15 billion annually. Annual growth rates for the total market are expected to be about 15 percent per year. The market can be broken down into three primary sectors: remediation services, end-of-pipe hazardous waste management, and pro-active source reduction. Each of these areas offers opportunities for construction firms.

Remediation services represent the most visible and rapidly growing area within the hazardous waste market. There are eight primary sources of projects in this area in the United States: federal EPA, state environmental agencies, DOE, DOD, private-party cleanups, RCRA corrective action, real estate development cleanups, and leaking underground storage tanks (LUSTs). Table 5.3 shows a breakdown of some of the major characteristics of each of these market segments (as developed by Andrew J. Hoffman in his master of science thesis for the MIT Department of Civil and Environmental Engineering).

As estimated by the *Environmental Business Journal,* a booming market is expected for remediation, with growth of perhaps 25 percent annually over the next 5 years, because much of the work under the Superfund program has been dedicated so far to planning and feasibility studies. Large numbers of these projects will be moving into the active phase over this period. Design work should also see healthy growth, despite the recession, along with the growth in treatment facilities and remediation. When *Engineering News Record* polled its readers about the trends in 15 different categories of the construction business for 1992, the hazardous waste market led the list with 70 percent of respondents expecting increased business. There may be slower growth in such hazardous waste sectors as transportation, storage, and disposal, however, as more of the industries generating these materials shift toward waste minimization, resource recovery, and on-site treatment. The market is more mature in the United States than elsewhere in the world, so construction firms should find growing potential overseas as foreign markets develop for hazardous waste cleanups.

There is obviously a need to prevent the creation of more abandoned hazard-

Table 5.3 Profile of U.S. Remediation Markets

Market	Distinctive Characteristics
Federal EPA	• 1,246 sites on EPA National Priority List (NPL) • 26,000 more sites on Hazardous Ranking System • GAO estimates up to 368,000 for a comprehensive inventory • Average cost of $20–30 million for cleanups to date • OTA estimates $500 billion will be needed over next 50 years to clean EPA NPL sites
State EPA	• 37 states had full fund and enforcement capabilities (1989) • 7 states had limited fund for emergencies • Collectively, states identified 50,000 sites: 28,192 may require cleanup and 6,169 are priorities • $415 million available for state Superfund cleanups
DOE	• In 1990, DOE identified 3,700 potential release sites and 5,000 "vicinity properties" that may be affected by DOE facilities • 17 DOE facilities on EPA NPL • $30 billion in DOE expenditures from 1991–95, 35 percent on remediation, 65 percent on waste management • Could need $150 billion total to clean all sites and bring facilities into full environmental compliance by 2019
DOD	• 4,401 sites that may need remediation • 96 DOD sites on EPA NPL • Spent $600 million on cleanups in 1990, up to $1.1 billion in 1991 • Complete restoration may cost $14 billion
Private-party cleanups	• Accounted for 60 percent of 1989 response actions at NPL sites • Growing interest to expedite cleanups before EPA involvement
RCRA corrective action	• Initiates cleanup of hazardous waste sites at operating facilities • 5,700 sites currently regulated under RCRA • Estimated total cleanup costs range from $7–42 billion
Real estate development	• Many states have developed laws that require environmental audits when: a property is sold, business ownership changes, companies merge, a company goes bankrupt, an industrial lease expires, or business operations cease. These audits seek to identify contamination or potential releases of contamination.
LUSTs	• EPA estimates 20 percent of underground storage tanks (USTs) may be leaking. These tanks include 2 million petroleum tanks and 50,000 hazardous substance tanks.

ous waste sites in the future. To address this need, RCRA now requires that companies implement effective cradle-to-grave management practices for their hazardous waste. Most estimates place the total annual generation of hazardous waste at between 250 and 500 million tons—a rate of some 1–2 tons for every

Table 5.4 Estimated National Generation of Industrial Hazardous Wastes

Major Industry	Estimated Quantity in 1983 (1,000 Metric Tons)	Percentage of Total
Chemicals and allied products	127,245	47.9
Primary metals	47,704	18.0
Petroleum and coal products	31,358	11.8
Fabricated metal products	25,364	9.6
Rubber and plastic products	14,600	5.5
Miscellaneous manufacturing	5,614	2.1
Nonelectrical machinery	4,859	1.8
Transportation equipment	2,977	1.1
Motor freight transportation	2,160	0.8
Electrical and electronic machinery	1,929	0.7
Wood preserving	1,739	0.7
Drum reconditioners	45	0.7
Total	265,594	100.0

person in the United States! Based on 1983 data, the largest generators appear to be the chemical, metal, and petroleum industries, as indicated in Table 5.4.

A distinct market for hazardous waste management services exists separately from remediation. This segment includes the transportation, storage, treatment, and/or disposal of hazardous waste. Since large generators of hazardous waste do much of their treatment on-site, the market for these services is far less than total expectitures on hazardous waste management. Booz, Allen & Hamilton estimated the market for hazardous waste management services at $12 billion in 1990, compared to $5 billion for remediation services. Generators of small quantities (100–1,000 kilograms [kg] per month) provided a waste management market of $1.5 billion in 1990, and demand is continuing to grow. Medical waste management offers an annual market of at least $1 billion, and the nuclear waste business is nearing a level of $1 billion a year.

Increasingly costly hazardous waste management is encouraging companies to avoid creating the waste in the first place. Just as in the solid waste sector, a waste-management hierarchy has begun to develop for hazardous wastes, starting with source reduction, then recycling, then treatment, and finally disposal as a final resort. Pollution minimization efforts have become widespread. Some voluntary initiatives include: the EPA's Industrial Toxics Program under which 600 U.S. companies agreed to reduce pollution from seventeen toxic chemicals by 33 percent by 1992 and 50 percent by 1995; the Chemical Manufacturers' Association's Responsible Care Program, with pollution prevention as one of its major tenets; and the Valdez Principles, a voluntary corporate ethics code drafted by the Coalition of Environmentally Responsible Economies (CERES), which covers a full range of environmental issues, from the wise use of natural resources to damage compensation.

Mandatory pollution-minimization programs are taking shape at the state level, although not federally. Massachusetts has perhaps the toughest law in the nation: a mandated goal of a 50-percent reduction in the use of toxics by 1997 from the 1991 level. Firms are beginning to recognize that pollution minimization is cost-effective. Although it is too early to quantify an actual market for technologies and services that minimize hazardous waste production, there appears to be an attractive potential for firms with capabilities in this area.

Even though the hazardous waste market presents the construction industry with perhaps the most tangible and quantifiable opportunities in the environmental marketplace, caution is in order. Liability insurance and bonding requirements (particularly for public work) are more difficult for contractors to obtain when operating in the hazardous waste business. In addition, comprehensive and somewhat labyrinthine regulations complicate work in this field.

Solid Waste

Solid waste disposal consists of four major activity sectors: landfilling, incineration (including waste-to-energy plants), recycling and composting, and waste minimization. Although they sometimes act as alternatives for each other, each activity should be viewed as part of an interrelated solid waste management system rather than as an independent whole. The different forms of waste largely determine appropriate disposal solutions. A hierarchical approach to a total solid waste management system starts with waste minimization, followed by recycling or composting of usable portions of the waste stream. Then the remainder might be incinerated, with any residual waste sent to a landfill.

Solid waste management is the largest and most mature segment of the total environmental marketplace, estimated to be about $31 billion in 1991. With a growth rate of about 8 percent a year, it is expected to reach $45.4 billion in five years. Considerable shifting will be taking place across the hierarchy during that period, however, opening up new business opportunities for the construction industry in some areas while other portions are in decline. As landfills either reach capacity or are closed down, and since few new ones will be put in place because of tougher construction requirements and siting problems, this portion may decline in volume. At the same time, however, higher tipping fees due to the shortage of landfill space will boost income for operators even as costs escalate.

The recycling sector is growing at a rapid pace as communities struggle to get some income from their solid waste while at the same time dealing with lost landfill space. Incinerators also face tough siting problems, particularly from local or regional NIMBY groups. Waste minimization efforts may cause a cutback on disposable packaging and an increase in the proportion of recyclable materials used. But the massive cutbacks in landfill space make it likely that there will still be large volumes of solid waste available for running waste-to-energy plants. These may gain favor because they can help reduce the need for fossil fuels and cut back on the requirements for new fossil-fuel or nuclear plants. To respond to citizen demands, however, new waste-to-energy plants may have to be located far from population centers, with facilities ensuring that toxic substances are filtered out of any emissions and removed from ashes that might be shipped to a landfill.

Even though waste minimization and materials recovery will help reduce the need for disposal, it is likely that we will still generate large amounts of waste in the future. Incineration and landfilling offer the best disposal methods avail-

Table 5.5 Solid Waste Disposal and Recovery by Mode in 1988 and Projected for 1995

	Millions of Tons		Percentage of Total	
	1988	1995	1988	1995
Total generation	179.6	199.8	100.0	100.0
Recovery for recycling	23.1	38.8	12.9	19.4
Recovery for composting	0.5	9.5	0.2	4.8
Total materials recovered	23.6	48.3	13.1	24.2
Discards after recovery	156.0	151.5	86.9	75.8
Combustion with energy recovery	24.5	45.0	13.6	22.5
Combustion without energy recovery	1.0	0.5	1.5	~0.0
Total combustion	25.5	45.5	14.2	22.8
Total landfilled	130.5	106.0	72.7	53.1

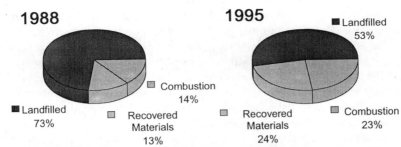

Figure 5.1 Solid Waste Disposal and Recovery by Mode in 1988 and Projected for 1995

Source: Franklin Associates, Ltd., *Characterization of Municipal Solid Waste in the United States: 1990 Update*, June 1990.

able from a technological standpoint, but difficulties will remain in finding suitable sites for these facilities. Projections for changes in the mix of solid waste recovery and disposal methods are summarized in Table 5.5 and Figure 5.1.

Several trends in solid waste management in the United States are highlighted by the data in Table 5.5 and Figure 5.1. Landfilling has traditionally been the preferred method of disposal because of its relative simplicity and lower cost. The EPA estimates that about 73 percent of municipal solid waste ended up in landfills in 1988. But alternative forms of waste management have begun to replace landfilling, and the trend has been accelerated as several states passed laws imposing mandatory source reduction, composting, or recycling. By 1995, the data shows that much larger amounts of solid waste will be recovered, and, of that not recovered, proportionately more waste will be incinerated. In 1988, some three-quarters of solid waste went to landfills, but this proportion is expected to drop to about half by 1995. The number of operating landfills continues to fall as stricter landfill regulations make it difficult for older facilities to continue operating, while siting of new landfills faces increasing restrictions and public opposition. What do these trends suggest might be the opportunities for construction in the solid waste area? Although the potential is not as quantifiable as in the hazardous waste sector, we will comment on construction-related potential in landfilling, incineration or waste-to-energy plants, recycling, and composting.

Landfilling can be a lucrative business for privately run facilities because there is frequently little or no competition within a region, even though pressures from regulators and public opposition groups have increased. Entry into the market is quite difficult because of much tougher regulations in recent years. Operating costs vary widely by region, but it is likely that in many cases tipping fees far exceed the operators' costs. A recent study in Michigan estimated costs for a modern, single-liner landfill. Using these figures, an analysis suggested a break-even price of $11.20 per ton of waste, assuming 1,000 tons a day for 320 days a year for 21 years, and using a discount rate of 10 percent. This calculation did not include insurance, taxes, bonding, corporate overhead, remediation, contingencies, or profit. Most tipping fees for comparable landfills far exceed the amount calculated. Average tipping fees for 11 counties in New Jersey, for example, are $66 per ton. This suggests significant profit potential for such facilities, even though landfill costs vary widely depending on location, and transport can add considerably to total costs.

Sanitary landfilling is a fairly straightforward procedure. In simple terms, earth is excavated and then, as solid waste is loaded into the excavation, successive layers are covered over with dirt in an operation called "cut and cover." Modern landfills include containment systems to block leachate and associated pollution that results from landfilling. Waste decomposition can form methane and other gaseous products, and equipment to remove these gases are required in new facilities. Unfortunately, the EPA estimates that about 90 percent of the waste now landfilled in the United States goes into dumps that lack containment systems. The failure of present landfills to control or collect these by-products has been a major source of criticism for this approach to waste disposal.

Wide recognition that indiscriminate dumping of mixed waste is the most environmentally unsatisfactory method of disposal has led to growing opposition to landfilling. Stricter regulations have made it more difficult both to continue to operate existing landfills and to find suitable sites for new ones. The NIMBY syndrome reflects general public opposition to siting landfills. In response, other means of waste management have gained in popularity, helping to divert at least part of the waste stream from landfills. Despite the negatives, landfilling will continue to play a major role (perhaps a dominant one in the United States) as part of the overall waste-management system. It is infeasible to recycle or burn all of the waste generated. And while waste generation continues to grow, landfill capacity is steadily shrinking.

To meet rapidly growing needs, more landfill capacity will be needed in the future either through enlarging present sites or construction of new facilities. New landfills will most likely require containment and collection systems to control waste by-products. Adequate technology does exist for such landfills, although construction methods are more complicated than for the simple excavations of the past. In spite of the opposition, demand for new high-tech landfills will certainly rise as capacity dwindles. Other countries with much less available land, such as Japan, face even greater restrictions on the use of landfills. Although construction companies may build some of these new facilities, large waste-management firms may also build their own. Even though the profit potential for operating a landfill may appear attractive to a construction firm, it must be recognized that tough regulations and wide public opposition make entry into this market extremely difficult.

Solid waste combustion *with* energy recovery is generally referred to as **incineration** or **waste-to-energy**. According to EPA estimates, by 1995 essentially all solid waste combustion will involve energy recovery, so we will refer to it here as "incineration" with the assumption that there is some form of energy recovery. Historically, interest in energy production from solid waste has increased when other energy sources (mainly fossil fuels) became more expensive. Energy prices climbed so high in the 1970s, in fact, that in some cases incineration actually became cost-competitive with landfilling. As energy prices have fallen since then, and regulations involved in building these plants have stiffened, they are much less economically attractive, although this situation may change again in the future as landfill requirements become much tougher and energy prices rise again.

Incineration has two other related advantages over landfills. Incinerators can be located closer to population centers if pollution is properly controlled, and they take up much less real estate. In countries with space constraints (such as Japan and many European nations), incineration plays a larger role in the overall waste-management system. In the United States, the 1970 amendments to the Clean Air Act required much more expensive pollution-control equipment for incineration facilities, adding to their costs. Subsequent clean air laws set even stricter laws for emissions.

Incineration continues to gain importance as part of the solid waste management mix around the world and even in the United States in spite of higher costs and growing public opposition. The same NIMBY syndrome and general public antagonism exists for incinerators as for landfills in the United States. Such opposition is not completely unfounded. There are potential problems with emissions of chemical toxins and heavy metals as well as with the reliability of some facilities, which could lead to excess emissions or a pileup of unburned trash. Incineration can also impede recycling efforts. Operators (sometimes municipalities) might oppose removal of paper and other recyclables from the waste stream because this might make their incineration facility uneconomic. Sometimes waste volume is part of a contract with an outside operator. At the same time, some advantages can be claimed for this approach, including energy production, reduction of solid waste volume, destruction of some pathogens and toxins, and, in some cases, the production of salable by-products. Incineration faces much less opposition in some other countries. There appears to be "little or no opposition to incineration in France," according to a recent article in *Civil Engineering,* which indicates that the same is true elsewhere in Europe. "Switzerland incinerates 80 percent of its waste, compared to 40 percent in France (27 percent with energy recovery, 13 percent without)."

Unlike the relatively simple technology of landfills, incinerators require high-technology design and construction. The Clean Air Act along with local emissions legislation regulate incinerators, and pollution-control equipment is a critical and expensive part of any facility. This makes technological sophistication along with high capital costs barriers to entry into the incinerator-construction business. More incinerators are being proposed by specialty construction firms that plan to finance, build, and operate the plants. But, because of the formidable technological requirments and capital risks, only large, well-capitalized firms with extensive technical experience are likely to pull off such ventures successfully. Since specialized technical knowledge is so critical, perhaps the best way for construction firms to enter this market would be through joint ventures with or acquisition of technology vendors. Like the landfill business, incineration does offer significant profit potential for those firms that can successfully enter the market. The downside of such ventures, however, is illustrated by Dravo Corporation of Pittsburgh, a construction firm that was forced out of business after a failed attempt to enter the incineration marketplace.

Recycling and composting are gaining in popularity as desirable and sometimes cost-effective alternatives within the solid waste management mix. Lawmakers, responding to citizen demand, have recently enacted various regulations aimed at removing recyclables and compostables from the waste stream. Analysis suggests, however, that oftentimes both practices may not prove cost-effective. Admittedly, the cost calculations do not account for hard-to-quantify factors such as the collective benefits (or costs) to the environment. How can the benefits versus costs of recycling be calculated, for example, when a recycling plant is driven by fossil fuel and the original producer uses renewable energy? Such evaluations are difficult to make. In spite of such drawbacks, however, recycling and composting do offer environmental benefits, and they will play a larger role in future waste management.

The primary problem facing recycling efforts is the lack of stable markets for recycled goods. Many organizations will buy only virgin materials because of concerns that recycled goods may be of inferior quality. Even those that *do* buy recycled products might switch away from them if prices decline for virgin materials. If secondary materials become acceptable resources for industry, these markets could develop in the future. An example of this trend in the aluminum industry is the plan of Alcoa to add only scrap-dependent facilities, using recovered materials, in the future. This makes good sense in the aluminum business where scrap-dependent plants cost about a tenth as much as conventional plants and can be built in about half the time. In cases where favorable conditions such as these exist, demand for recycling facilities will continue to increase. Other pressures should broaden the trend toward recycling of more materials, including higher costs for landfilling, community encouragement of citizen recycling, and legislative mandates for greater use of recycled materials in packaging and other products. This will create opportunities for construction firms. Some recycling facilities are highly mechanized and others are low-tech and labor-intensive. Currently in the United States, each captures about half the market. The mechanized facilities typically require higher capital investment but have lower operating costs that their low-tech counterparts. As more sophisticated plants are built, construction firms with technical experience should gain an edge, just as occurred in highly specialized incinerator technology.

Even though composting is the least common method for dealing with solid waste, it is gaining in importance in some areas because it complements other waste-management programs. The primary problem with composting involves setting standards for types and quality of compost. For markets to reach their full potential, compost must become a commodity that has standard and reliable integrity and content. Any biodegradable substance is compostable. This includes such substances as food waste, yard waste, wood, and paper; and Franklin Associates estimates that such substances make up 65 percent of the waste stream. Potential markets for compost exist in the nursery and greenhouse businesses, where large demand exists for organics to make good-quality soil. Costs for compost plants vary widely with location. Capital costs for new facilities ranged from $5,000 to $83,300 per ton per day of capacity, while operating costs ranged from $9 to $85 per ton, according to an industry survey. Construction demand should rise for composting facilities just as for recycling plants in the future, although it is difficult to make accurate monetary projections.

Waste Minimization

Waste minimization offers the ideal solution for solid waste, just as it does for hazardous waste. The benefits of generating less waste flow to everyone. Producers save by cutting waste, because it both reduces the cost of input to production processes and lowers cost of waste disposal. Consumers benefit by paying lower prices for more efficiently made products and because they will not have to dispose of so much trash. Unfortunately, achieving this objective is not a simple matter. It is difficult to induce citizens to minimize waste because the cost of waste disposal may be negligible to them. Waste disposal cost is built into property taxes in most communities, so there is no difference in cost whether an individual disposes of one or fifteen bags of trash per week. Firms often shy away from adopting waste-minimization technologies because they require immediate capital investments that pay dividends only in the long run. Under the present system in the United States, managers tend to avoid expenditures that cut into short-term profits even though the result might be greater long-term efficiency .

Despite the difficulties of implementing waste-minimization policies, the concept fits well into the sustainable-development concept now gaining wide credence. It provides benefits for the environment while at the same time offering business advantages. Waste minimization is part of the "total quality" creed now sweeping industry around the world. Total quality goes beyond just producing quality products—it requires a broader view of achieving quality in all operations, and part of this view is to minimize waste. Japanese companies have demonstrated that achieving total quality is not more expensive, difficult, or time-consuming. They view it instead as necessary to remain competitive, and they see that it returns more than it costs if it is consistently designed into the manufacturing process. Many countries are implementing waste-minimization policies, and some companies in the United States have developed their own programs emphasizing reducing waste, reusing materials such as solvents, and finding markets for scrap. The market for construction is ill defined and difficult to quantify in this sector, but industry will be looking to construction contractors to help them meet waste-minimization objectives in the future, so there should be ample opportunity for those firms capable of making a contribution.

Sewer / Wastewater

The market for wastewater treatment by industry and communities was estimated to exceed $15 billion in 1991, and an annual growth rate of 12 percent is forecasted for the next few years in the United States. This industry sector's growth accelerated after the passage of the Federal Water Pollution Control Act (FWPCA) in 1972. That legislation resulted from growing concern over water quality, fueled particularly by stories in the public press, such as a *National Geographic* magazine report on high pollution levels in the Great Lakes. Since the waters of our streams, rivers, lakes, aquifers, and coastal seas help support life, they are among our most treasured national assets. Yet there was increasing evidence that the regulatory framework existing up to that time was failing to keep the waters clean. Earlier water pollution legislation had created a municipal-grants program for building sewage facilities, established federal enforcement authority, and given states the authority to set and enforce their own water-quality standards. FWPCA was a major revision of these earlier laws, strengthening the municipal-grants program and adding clout to the federal role by authorizing the EPA to issue point discharge permits.

The 1972 law ensured that cities could receive help in building sewage treatment facilities by setting up a massive federal construction grants program. Both municipal and industrial sources of wastewaters were directed to treat discharges before they poured into waterways. A National Pollutant Discharge Elimination System (NPDES) gave the EPA authority to set pollutant limits for point sources of effluents by means of discharge permits. Although industrial facilities that discharge wastewater into sewage systems are not covered by NPDES permits, they must meet other Publicly Owned Treatment Works (POTW) standards. The law allowed states to retain the authority to impose water quality standards even more stringent than federal levels when deemed necessary. The 1972 law was also the first to offer limited protection to wetlands, an issue currently being debated as part of the proposed reauthorization of the Clean Water Act.

The underlying goal of the new legislation was eventually to eliminate all discharges of pollutants into U.S. waters. This legislative mandate was further strengthened with the Clean Water Act of 1977, and the impact was highly beneficial. Water pollution from point sources was significantly reduced. Recently, in fact, the EPA claimed that "less than 15 percent of the nation's remaining water quality problems stem from point sources, especially industrial dischargers." While great progress was made in cleaning up effluents from industry, however, it became apparent that contamination from more distributed sources, such as runoff from farms and from urban areas, including untreated discharges from storm sewers, was increasing. This knowledge led to the Water Quality Act of 1987 (also called the Clean Water Act), which began to address the problems of non-point sources of pollution. The 1987 legislation also began to shift more of the responsibility for financing construction of sewage treatment facilities from the federal government to states and localities.

Regulatory pressures under these laws has generated huge expenditures for water and sewage treatment in the United States over the past two decades. The EPA estimates that $590 billion (in 1990 dollars) has been spent on water pollution control in the United States since the 1972 revision went into effect. Surface waters have been made substantially cleaner as a result. But the original goal of zero discharge of pollutants has yet to be realized.

The next stage in the evolution toward cleaner waters will be the renewal of the Clean Water Act, which will be required because many of the programs authorized under its provisions are running out of funds. A Senate bill has been introduced and bills relating to clean water are also taking shape in the House. Although other legislation may overshadow these bills, politics is likely to revive concerns over pollution issues when election time approaches. Political debate will undoubtedly focus on increasing federal funding for construction of more wastewater treatment facilities, methods for controlling non-point sources of pollution, and wetlands protection.

Water Supply

The $12 billion estimate for the water utilities market segment by *Environmental Business Journal* may be somewhat high due to the lingering recession, and the slowdown will likely continue well into 1992. But there are some signs that activity will pick up soon, especially for drinking-water treatment, due to new federal requirements. In fact, because of stiffening requirements, the low 4-percent growth rate projected for this sector over the next five years may turn out to be too modest. The reason is an existing EPA order that requires suppliers to filter surface waters by June 1993 unless exemptions are granted by state officials. Many utilities have delayed taking action, however, while awaiting two EPA rulings that will impact future upgrading of treatment facilities. One ruling pertains to the by-products of disinfection, and the other concerns radon regulation. Once these rules are announced, numerous projects now in the planning stage are likely to move toward implementation. Many small systems that draw on groundwater "are not used to treating water at all," claims Jack Hoffbuhr, deputy executive director of the American Water Works Association. "In the next five years we're going to see an awful lot of activity," he adds. Demand may also pick up for desalinization plants. Coastal areas in the Southeast and West will face problems of shrinking water supplies along with increasing contamination problems in the next few years, so they are expected to turn to desalted seawater to supplement supplies.

Although specific figures for water-control projects are not included in the market-size and growth-rate figures presented here, some comments on the outlook are in order. River-control projects, such as dams for flood control and power generation, face an uncertain future, according to a 1992 forecast by *Engineering News Record* (January 27, 1992), because of a combination of financial and regulatory constraints as well as environmental requirements and opposition by conservationists. By contrast, a brighter future is forecast for waterway construction and improvement projects, mainly for transportation.

Air Pollution Abatement

Air pollution control is a relatively small market today at about $6.3 billion annually, but it appears to be on the brink of very rapid growth. The reason: the passage of the new Clean Air Act, the most influential piece of environmental legislation in recent years because it expands regulation to include much tighter control of toxic gases and also applies the tradable-permit concept to the air pollution sector at the federal level. The provisions of this far-reaching bill will also significantly influence the energy and petrochemical industries, which we will discuss in later sections.

Renewed concern over sulfur and nitrogen oxides, which have been blamed for acid rain, was a driving force in the reauthorization of the Clean Air Act in November 1990. But there was another major thrust in this bill toward reducing air toxics, or hazardous air pollutants. The legislation specifically targets 189 substances, classified as air toxics, for emissions reduction. These air toxics include heavy metals and organic compounds that are suspected of posing risks for human health and the environment. While emissions of suspected acid rain precursors such as sulfur dioxide and nitrogen oxides, along with flue-gas particulates, have been heavily regulated in the past, there was much less stringent regulation of air toxics. The new legislation gives the EPA far more authority for regulating these hazardous emissions. Within industries that are considered significant emitters of air toxics, the EPA will identify sources of emissions and require the companies that exceed minimum limits to employ maximum achievable control technology. The limits for each source are emissions of 10 or more tons per year of any one of the 189 listed substances, or any combination of them adding up to 25 tons per year.

Industries most likely to be immediately affected by these rules are petrochemical processors and metal producers. Measurement problems will allow the utilities industry to escape immediate regulation of air toxics even though 37 of the 189 listed substances are known to be emitted by power plants. There are problems with pinning down the exact constituents of these emissions and any associated risks are not easily quantified. Trace amounts of toxics are known to exist in coal, which is widely used to generate electric power, but the trace levels vary widely. Many differences exist in the design and operation of power plants, and the complexity of sampling and analyzing trace pollutants from each source adds to the difficulties.

Because of these uncertainties, the 1990 amendments to the Clean Air Act calls for the EPA to undertake a three-year study of any health risks from air toxics emitted by utilities. It also mandates a four-year study of the effects of mercury emissions from utilities and other sources. The National Institute of Environmental Health Sciences will conduct a separate study to define the threshold of mercury exposure that may adversely effect human health. Until these studies are completed, the EPA has been directed to put off air-toxic regulation of the electric utility industry. What market impact can be expected if utilities are added to the list of those that must comply with the air-toxic provisions of the 1990 Clean Air Act amendments? Estimates put the costs at about $7.8 billion a year on top of the cost of complying with provisions to reduce acid-rain emissions.

REGULATORY IMPACTS ON ENERGY AND PETROCHEMICAL INDUSTRIES

The following two sections on energy and petrochemicals deal with trends in two industries that will undoubtedly be most heavily impacted by growing environmental concern and increased regulation. These industries are likely to require substantial investment in construction—both in retrofitting existing plants and in new facilities—to meet a steadily growing list of environmental restrictions. Since both industries play such a critical role in the environmental picture and their facilities are often huge—major power plants, refineries, or petrochemical processing plants—the work they need done may create some of the largest construction projects. Even though handling such jobs requires the project-management skills and wide-ranging capabilities of major construction contractors, there will also be growing opportunities for smaller, specialized vendors to make contributions as part of a project team. Since even large construction firms often do not have experience in areas such as environmental controls and treatment, these segments offer particular targets of opportunity.

Energy

Our discussion of the energy sector will focus on electric power production rather than on other uses of energy, for transportation or heating, for example. Modern societies depend heavily on energy supplies for increasing their standards of living. As population levels grow, more energy is needed to provide them not just with more food and goods, but also with productive jobs so that individuals can provide for their needs. Greater wealth creates demands for richer lives, including more opulent goods and other amenities, such as travel and leisure activities. All these trends lead to greater use of energy. Economic expansion was long believed to require proportional increases in energy supplies, and it has been in only the last couple of decades that it has become clear that this is not an immutable relationship. An industrial economy's ability to expand without concomitant increases in energy was demonstrated in the United States following the oil shocks and higher prices of the 1970s. Because of higher costs for energy, conservation efforts rippled throughout the American economy, from more fuel-efficient buildings, aircraft, and cars, to higher-efficiency machines, appliances, and lighting, as well as greater recycling of glass and aluminum. Growth in energy production became "decoupled" from the expansion in the economy because of these changes in practices. More recently, as oil has become more plentiful and prices have declined again, the trend toward conservation has slowed down. Even before the oil shocks, higher fuel prices in European countries and Japan, partly due to taxation, had encouraged a higher degree of conservation, and their economies currently use about half the energy per unit of GNP as the United States.

The ability of modern economies and populations to grow without commensurate increases in energy is vital to the concept of sustainable development because energy production and consumption are strong contributors to many forms of environmental degradation. Tapping fossil fuels sometimes pollutes the oceans through oil spills or seepage from seafloor wells, while waste from coal mining fouls rivers and streams. Emissions from industry and power-plant smokestacks contribute to urban smog, acid rain, and greenhouse gases. Thousands have become ill and many have died from the effects of radioactive emissions from the accident at Chernobyl's nuclear plant. Even hydroelectric dams contribute to environmental degradation by killing fish and flooding millions of acres of farmland and living space.

Because so many of today's environmental concerns are tied into the burning of fossil fuels, the electric utility industry is being pressed to cut emissions while at the same time producing more power. As we said earlier, a study by MIT's Energy Laboratory indicates that many utilities could cover the capital costs for modernized plants with the savings on fuel costs produced by switching to much more fuel-efficient technology. Burning less fuel for an equivalent amount of energy production would simultaneously cut emissions, so it could cut these utilities' costs for pollution control systems as well. If little refurbishing is done, the industry will still have to spend significant amounts on equipment to control sulfur and nitrogen compounds and particulates under the Clean Air Act amendments. But if sentiment in the industry turns toward upgrading facilities to meet tighter regulations rather than spending more on pollution controls for aging plants, considerable construction activity could result for new power-generation equipment.

A vexing environmental problem is the contribution of fossil-fuel consumption to the buildup of greenhouse gases in the atmosphere. This has sparked a major controversy, with widespread disagreement over scientific, economic, social, and political aspects. Prevention of global warming was one of the main topics on the agenda for the Earth Summit held in Rio de Janeiro, Brazil, during the summer of 1992. Discussion centered on what can be done to curb the emission of greenhouse gases—carbon dioxide in particular. Although there are technologies to deal with sulfur and nitrogen compounds, there are no good technological fixes for reducing carbon dioxide efficiently. In fact, some environmental measures, such as scrubbing coal gases to remove sulfur dioxide, actually increase the amount of carbon dioxide released per unit of energy production. Whatever route is chosen to deal with this problem in the future, it is certain to have a significant impact on construction plans for new facilities, and it will certainly create a demand for modifying present processing facilities.

Carbon dioxide accounts for about half of the greenhouse gases in the atmosphere, and the United States currently produces some 20 percent of the world's total carbon dioxide. While many industrialized countries have pushed for a freeze or reduction in carbon dioxide emissions, the United States has resisted setting fixed targets. The major reason given for the U.S. position has been the potential economic impact of curbing emissions. Some studies support this position, but others suggest that efforts to reduce greenhouse gases could actually be profitable in the long run. This position has drawn the support of environmentalists and some government analysts, business leaders, and economists. These advocates believe that efforts to reduce emissions would force more efficient uses of energy, thus freeing up large amounts of capital for productive investment and making American companies more competitive internationally.

Environmental groups urge that carbon dioxide emissions be cut mainly through energy conservation. Although persuasive arguments can be made for

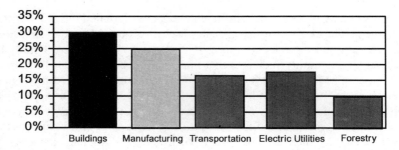

OTA Estimate of Where Cuts in Carbon Emissions Would Come From if the U.S. Reduced Total Annual Carbon Emissions 35% by 2015. (Represents 1987 Level)
Note: Interpret graph as follows: 30% of the total 35% cut in carbon emissions comes from the Building Sector.
Figure 5.2. Percentage Reductions in Carbon Dioxide Emissions (of total) from Five Economic Sectors

conservation as the most logical approach, in practical terms it is unlikely to be chosen as the sole solution. The amount of fossil fuels burned could be reduced by a wide range of conservation measures, including more efficient power plants and factories, energy-efficient buildings, windows that cut down on heating and air-conditioning requirements, and more fuel-efficient cars and aircraft. Energy needed for industrial processes could be reduced by greater recycling of scrap and other by-products, particularly energy-hogging materials such as aluminum. Some of these steps would require considerable capital investment and the gradual replacement of existing technology, (large cars and poorly insulated buildings, for example). Practical implementation of alternatives such as these would likely take decades.

OTA recently estimated that, if no special action is taken, carbon dioxide emissions will rise about 50 percent over the next twenty-five years. OTA then developed two potential long-term paths to reducing carbon dioxide emissions: a "moderate" scenerio allowing emissions in 2015 to be 15 percent *higher* than those in 1987, and a "tough" scenerio in which emissions would be 35 percent *lower* in 2015 than in 1987. OTA then calculated how much carbon dioxide reduction would be needed within five economic sectors: buildings, transportation, manufacturing, electricity generation, and forestry. For the tough scenerio, the total reduction would be split among the five sectors as shown in Figure 5.2—30 percent from commercial and residential buildings, 25 percent from manufacturing, between 15 and 20 percent each from transportation and electric utilities, and 10 percent from forestry.

The projected reductions under the moderate scenerio would be split similarly, except that forestry would play a somewhat reduced role. Considerable construction activity would be necessary in three of these sectors—buildings, manufacturing, and electrical utilities—to achieve the proposed carbon dioxide emission reductions.

The buildings sector offers the greatest potential for emission reduction, but it also might prove the most difficult for implementation. The OTA suggested that there are a multitude of areas where energy consumption, and associated emissions, could be reduced, ranging from design and construction to furnishings such as lights and appliances. For new buildings, the OTA sees opportunities in better insulation, tighter windows, and improved construction methods that could lower heating and cooling needs. Existing buildings offer less potential for reducing emissions, but some steps could be taken, such as installing better heating/cooling systems, more efficient appliances, and energy-efficient windows. The OTA sees a greater challenge in forging the policies to bring about such improvements than in actually applying technological fixes. Recommended policies should be directed at allowing profitable demand-side management by utilities, initiating technology-specific regulations such as appliance standards and energy codes for buildings, and providing better consumer information on ways to conserve energy.

Three technical improvements were identified by the OTA as holding the most promise for reducing carbon dioxide emissions in the manufacturing sector. These improvements are: process changes that would allow more energy-efficient production; cogeneration of electricity and steam for industrial processes, which would make more efficient use of fossil fuels; and more efficient motors. Potential policy instruments for inducing such improvements in energy efficiency include carbon taxes, marketable permits, and specific efficiency standards. Efforts should be concentrated on the four industries that use over 75 percent of the total energy consumed in the manufacturing sector: paper, chemicals, petroleum, and primary metals.

Currently, electricity generation accounts for about a third of the carbon emissions in the United States, and the OTA predicts that, if nothing is done, this share may rise to 45 percent by 2015. OTA cited two *moderate* measures for achieving reductions: improving the efficiency of fossil-fuel-fired plants through better maintenance; and operating existing nuclear plants 70 percent of the time instead of the current 60 percent and extending their useful lives from thirty-five to forty years. *Tough* measures would focus on shifting demand for energy away from coal. Included would be requirements that all new plants use renewable energy sources, improved designs for nuclear plants, and the use of high-efficiency gas turbines. In addition, existing fossil-fuel plants would be retired after 40 years instead of the typical 60 years of operation. If all the strict measures were put into effect, the OTA estimates that carbon emissions would be cut 14 percent from current levels by 2015. Potential policies for inducing these changes include: carbon taxes or marketable permits; emissions limits and efficiency standards; and federal funding of R&D and demonstration projects of renewable energy, conservation, and nuclear energy.

Another option for reducing the burning of fossil fuels is greater use of alternative energy sources such as windmills, photovoltaics, and geothermal power. So far, even with considerable subsidization of alternate energies during the 1970s, particularly by California, their costs have not yet come down enough to match the low costs of fossil fuels. But research programs and efforts to create market incentives during that era did lead to dramatic progress toward lower costs, so that a revival of more active support might help bring prices down even further. Already it appears that wind power is becoming competitive in some areas of the world, and the result could be a boom in demand for windmill construction.

Supporters of greater use of alternative energy sources complain that today's prices for fossil fuels do not take into account the full costs to society, which gives these fuels an unfair advantage in the marketplace. What about the huge health costs and lost work days stemming from respiratory and other disorders caused by breathing tainted air? Such health effects should be considered as part of the "true cost" of burning fossil fuels, contend California policymakers. They cite such costs to society to justify tough regulations forcing reduced auto emissions and less burning of fossil fuels.

Currently, fossil fuels are used to produce about two-thirds of the world's electricity. To achieve significant reductions in carbon dioxide emissions would

Table 5.6 World Electricity Production in 1986 (10^9 Kwh)

Region	Thermal	Hydro	Nuclear	Geothermal	Total
North America	2,059	642	485	15.00	3,202
South America	83	281	6	0.00	370
Europe	2,621	664	832	6.00	4,128
Africa	181	49	4	0.30	234
Mideast	170	22	0	0.04	191
Far East	1,115	326	229	6.00	1,676
Oceania	120	39	0	2.00	161
World Total	6,349	2,027	1,556	30.00	9,962
Percentage	63.7	20.3	15.6	0.30	100

Source: Darkazalli, "Energy and the Enivironment in the 21st Century," *MIT Press*, 1990.

Table 5.7   U.S. Electricity by Fuel in 1987

Coal	Petroleum	Natural Gas	Nuclear	Hydro	Geothermal
56.9%	4.6%	10.6%	17.7%	9.7%	0.5%

Source: Darkazalli, "Energy and the Environment in the 21st Century," *MIT Press*, 1990.

require substantial changes to the existing system for generating electricity. The mix of methods for generating electricity in different parts of the world is illustrated in Table 5.6. The mix of methods for the United States is shown in Table 5.7.

Efforts to curb emissions of carbon dioxide and other greenhouse gases would certainly require major construction efforts. Many electrical power plants that use fossil fuels would need to be retrofitted with more efficient generating facilities, and many new power plants may need to be built. Since currently all noncarbon-based energy-generating sources—including hydro, nuclear, solar, wind, geothermal, and photovoltaic—are used to produce electricity, an increase in the use of these sources would mean an expansion of total electrical generating capacity. Unfortunately, all of these alternate energy sources currently involve higher capital costs than fossil-fuel plants, but lower costs for primary energy inputs would help offset the incrementally greater investments. A study done of the New England electrical system concluded that savings in operating costs might exceed investments in more efficient generating equipment in existing plants. Policies requiring more efficient energy production would create a need for new technology, and the replacement of older equipment and plants would create expanded markets for the construction industry. Experience in installing new technology could also lead to significatly expanded construction work to meet new demands for electricity in both the developed and developing world.

Petrochemicals

This major industry outputs a multitude of products, including petroleum for use as fuel as well as petrochemicals used for a wide range of materials including plastics, solvents, and synthetic fibers for textiles. Like the energy sector, the petroleum industry has traditionally been a megaproject customer for construction companies. Extraction facilities, refineries, and petrochemical plants represent some of the largest construction projects ever undertaken. But these plants are also potentially large polluters; therefore, they have traditionally been a major market for emissions-control equipment and fluid-treatment systems.

As regulation stiffens for acid-rain precursors and greenhouse gases, these facilities will require upgrading.

Since refineries produce the fuels used for transportation and a large proportion of power production (most of the 43–44 percent not provided by coal), any changes in fuel specifications to meet environmental goals will require modifications of refinery processing systems. Construction firms hoping to participate in this massive changeover, however, must remain alert to the ongoing national debate centered on the issue. Actually two changes in how gasoline is made are proposed in the 1990 amendments to the Clean Air Act—a relatively simple reformulation in 1995 to lower the vapor pressure of gasoline, and a much more costly revision for the year 2000. If the rules set up by the latest Clean Air Act provisions are not revised, all gasoline refineries would require retrofitting by early in the next century. The oil companies are urging reconsideration of this approach because, the industry claims, similar results could be achieved at much less cost to consumers if a canister were built into all automobiles. This canister would trap unburned gasoline vapor so it could go through an additional combustion cycle, thus eliminating excess exhaust gases. Automakers oppose this approach because it might add some $500 to the cost of a new car. But the refiners counter that this would still be far less than the additional cost of gasoline over the life of a car if refineries have to be redesigned. If the refiners get their way, an additional problem will be created—the disposal of the carbon adsorption canisters from junked cars, which would probably be designated as hazardous waste by the EPA. A better approach than either redesigning refineries or adding canisters to vehicles might be to develop engines that burn fuel more efficiently and thus avoid creating so much pollution in the first place, but neither industry is offering this approach as a practical alternative. As environmental rules continue to tighten, there is bound to be greater contention like this between powerful petroleum refining and auto manufacturing interests.

While environmental pressures increase on the U.S. petrochemicals industry, the big oil companies are also feeling a financial pinch, particularly in domestic markets. In 1991 net profits for the top oil companies' foreign operations fell only 2 percent, but domestic earnings declined sharply, down 36 percent. Costs for producing and refining oil are going up in the United States because of new environmental laws. At the same time, the industry is facing patchy demand and declining prices for its products. Many oil producers have been directing more of their resources toward foreign markets, and refiners also may begin to shift abroad. Because there has been more exploration for oil and gas in U.S. territory than in any other region of the world, analysts have expected that eventually many operations would move overseas. But the uncertainty and bureaucracy associated with U.S. environmental demands appear to be accelerating the trend.

WORLDWIDE ENVIRONMENTAL MARKET TRENDS

The United States has been among the leaders of the industrialized nations in percentage of GDP spent on pollution control. In the mid-1980s, the United States was estimated to be spending slightly less than 1.5 percent of GDP on pollution abatement and control, second only to West Germany, which spent about 1.5 percent. By 1991, it is estimated that U.S. spending had risen to about 2.2 percent of GDP, and the EPA expects the percentage to continue to increase to some 2.7 percent of GNP by 2000. Although Japan's capital spending on pollution abatement and control was lower in the mid-1980s at about 1.2 percent of GDP, the federal government in Japan spent more on controlling pollution than the government in any other country, while Japanese industry spent very little in this area. A quick overview of the market outlook in other areas of the world reveals some important differences in comparison with the United States. Figure 5.3 shows expenditures on the environment by several indusrial countries as a percent of their GDP.

European Markets

Observers often compare Europe's current environmental market with that in the United States in the early 1980s, before more stringent rules emerged as a result of legislation such as RCRA. The European marketplace can be broken into three distinct divisions. The first division includes countries that have developed comprehensive environmental policies over the past two decades and

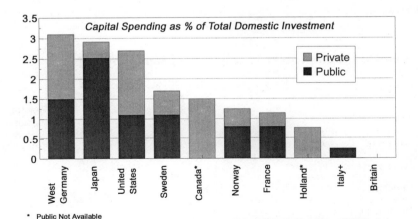

Figure 5.3. Investment in Pollution Abatement and Control in the Mid-1980s

therefore have well-defined markets. This group includes West Germany, Denmark, and the Netherlands. In the near term, only modest market growth is expected for environmental goods and services in these countries.

The second division has developed less rigorous environmental policies, and these countries typically also have less effective administrative and enforcement structures. Included in this group are Belgium, France, Ireland, Italy, Luxembourg, and the United Kingdom. A higher rate of near-term growth is expected within this division.

The third division consists of countries with the least advanced environmental policies, including three southern European countries: Spain, Portugal, and Greece. Major efforts will be required by these nations to meet legislative requirements of the EC, and, as a result, they are expected to have the highest growth rates for environmental markets. Estimated dollar figures and per annum growth rates between 1990 and 1995 are detailed in Tables 5.8 and 5.9 taken from *Environmental Business Journal,* March 1991.

While the fastest growth rate, 17.8 percent, is expected for the contaminated land remediation sector, it is also far smaller than the other groups in total dollar figures (Table 5.8). Similarly, although the countries in Division 3 have very high growth rates, their total markets are dwarfed by the magnitude of the West German, French, and United Kingdom markets (Table 5.9).

In the air pollution market segment, significant growth is expected for controlling industrial emissions of gases and vapors, especially those contributing to acid rain, and for cutting vehicle emissions. New directives on municipal wastewater and sludge are expected to spur growth in the water treatment industry. Remediation of contaminated land is growing as fears rise about health risks. So far, just as was the case in the United States as remediation got under-

Table 5.8 European Pollution-Control Market by Segment

Market Segment	Market Value ($ Billions)		Real Growth % per Annum
	1990	1995	
Air pollution Control	12.5	16.5	5.8
Water/wastewater treatment	16.6	27.5	10.7
Contaminated land remediation	1.3	3.0	17.8
Waste management	27.1	36.1	5.9
Totals	57.5	83.1	7.7

Source: Ecotec, from *Environmental Business Journal,* March 1991.

Table 5.9 European Pollution-Control Market by Country

Country	Market Value ($ Billions) 1990	Market Value ($ Billions) 1995	Real Growth % per Annum
Division 1:			
Denmark	1.0	1.4	7.0
Germany	21.9	28.1	5.1
The Netherlands	3.3	4.3	5.3
Division 2:			
Belgium/Luxembourg	1.2	1.8	8.4
France	9.8	14.5	8.0
Ireland	0.4	0.6	8.3
Italy	6.4	9.7	8.7
United Kingdom	10.5	17.5	10.6
Division 3:			
Greece	0.4	0.7	14.1
Portugal	0.5	1.0	13.5
Spain	2.1	3.8	12.7
Totals	57.5	83.4	7.7

Source: Ecotec, as shown in *Environmental Business Journal,* March 1991

way, most early site cleanups have involved excavation and removal of contaminated soil. But recently a growing market has been emerging for on-site and in-situ cleanup methods. Although the EC is examining mechanisms for funding such cleanups, it is not considered likely that they will take the form of an environmental liability system like that in the United States. New directives also expected for hazardous waste management will most likely lead to increased pretreatment and incineration of wastes.

As the EC continues to integrate economically over the decade, greater harmonization of environmental standards are expected across the community. While the European market will continue to grow, there may be significant differences among countries and sectors. Two groups are likely to merge, according to Dr. Haines: those nations that aim to introduce more stringent regulations ahead of the EC and those that simply strive to meet whatever standards are set for the community as a whole. He suggests that "as standards in the EC approach those in the United States, American companies, many of which lead their European counterparts with respect to experience and technology, will see the market for their pollution control products and services continue to expand."

Another study of European environmental markets concludes that spending by EC governments, companies, and individuals on environmental projects will be so substantial over the coming decade that the levels will rival those in the United States by 2000, but the mix is likely to be quite different. Regional differences in needs and priorities will make "pan-European" strategies difficult for outside companies to implement, conclude Bradford S. Gentry and James M. Shurtz of Britain's Morrison and Foerster in their report, "Developments in the European Environmental Market: Opportunities and Constraints," October 1991. Although there has been considerable comment about the appalling environmental conditions in Eastern Europe, these authors point out that conditions in parts of Western Europe are not much better. Belgium has problems with sewage, heavy metals, and nitrates in its rivers, while the Netherlands and Greece also suffer sewage problems. Milan dumps an excess of sewage into the Po River in Italy and is causing massive pollution in the northern portion of the Adriatic Sea. Stiffening environmental rules and enforcement in response to public outcry will spur markets for equipment and services all over Europe, although the emphasis will vary in different regions. Some Western European companies are already well positioned in the market and are working hard to

Table 5.10 Estimated Expenditure (Pounds, Billions), 1991–2000

Environmental Area	United Kingdom	European Community	United States
Greenhouse effect	48	237	443
Water quality	25	75–100	71
Waste management	19	180–200	120–170
Acid rain	11	51	25
Heavy metals	9	80	52
Ozone depletion	7	70	76
Air quality	7	34	17
Noise	6	32	33
VOCs & smells	3	26	27
Persistent organics	2	23	16
Contaminated land	2	25	150
Major spills	1	7	7
Totals	140	860	1,060

Source: Centre for Exploitation of Science and Tech. (UK) as shown by Gentry and Shurtz.

WORLDWIDE ENVIRONMENTAL MARKET TRENDS

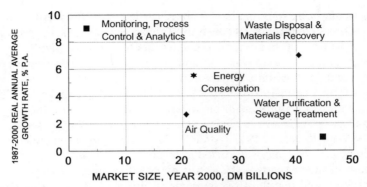

Figure 5.4. Market Size and Growth, Western Europe: 1987–2000
Source: Helmut Kaiser, as shown by Gentry and Shurtz.

capture available opportunities they have found. But some companies in the United States and Japan, in particular, have developed experience and technologies that could be a basis for developing relationships with local firms. With European spending steadily rising and so many companies already involved, there are likely to be few "quick hit" projects available for companies outside the Common Market, but these analysts still believe there will clearly be major opportunities for discerning investors.

Gentry and Shurtz made projections of estimated expenditures (in British pounds) for a wide range of environmental sectors over the period from 1991–2000 in the United Kingdom, the EC as a whole, and the United States (see Table 5.10.)

While there is little difference in estimates of spending in the EC and the United States for some environmental sectors, such as noise, VOCs and smells, and ozone depletion, there are some sectors where expenditures from 1991 to 2000 are expected to be much greater here than in Europe, such as waste management, acid rain, and heavy metals. Spending in the United States is expected to be much higher than in the EC for the greenhouse effect (because of higher reliance on fossil fuels for energy) and for remediation of contaminated land. Gentry and Shurtz also indicated estimates of Western European market sizes and growth rates as projected by Helmut Kaiser for the period 1987–2000 as shown in Figure 5.4.

Pacific Rim Markets

Rapid economic development in the Pacific Rim region, particularly in Southeast Asia, has brought with it growing environmental problems. As a result, environmental markets in this area of the world are receiving increased attention, particularly as some nations there reach levels of prosperity that enable them to address environmental concerns. Taiwan appears to be the most promising market at present, but other nations may soon follow its lead.

Taiwan has set up a timetable to address its major environmental problems by 1997, and that nation's EPA estimates that the task will cost some $37 billion over the period 1991–1997. That total breaks down into $3.5 billion for consulting and $17.5 billion for engineering services, $13.6 billion for equipment, and the remainder for testing and analysis, according to *Environmental Business Journal,* March 1991. After 1997, once the environmental infrastructure is in place, spending is expected to level off.

To gain outside expertise, Taiwan is expected to import up to two-thirds of the consulting and one-third of the engineering services. Since many of Taiwan's regulations are modeled on American standards, U.S. consulting firms have been involved in many early projects, and there is usually little modification required for either U.S. equipment or services in the Taiwanese market. But the Taiwan EPA, which is largely patterned after the U.S. agency, now requires foreign firms to team up with local businesses to get work there, thus increasing the need for joint ventures.

Regulatory activity has increased and environmental expenditures doubled recently in all the members of the Association of Southeast Asian Nations (ASEAN), the organization reports. Singapore and Malaysia are two of the most promising markets. While the largest environmental market in the region is in Japan, the Japanese market, unfortunately, remains very tough to penetrate. Japanese industry, with guidance from the government, is very reluctant to do business with outsiders, and this seems to be especially true in the environmental market.

Other areas of the world have significant potential for developing environmental markets, but weak economies and unstable political milieus may limit viable opportunities. Environmental needs are great in Eastern Europe and in the countries of the former Soviet Union. As these regions begin to deal with the World Bank and become members of the International Monetary Fund (IMF), their economies may begin to grow so that they can afford to address the serious environmental problems that were largely ignored when they were parts of the Soviet bloc. European countries, particularly the former West Germany, are in the best position to take advantage of these emerging marketplaces. But American firms with advanced skills or technology might be able to collaborate with Western European firms as these environmental markets develop. Other areas, such as Latin America and the Middle East, have significant environmental problems, but economic weakness and political strife make doing business there problematical.

In conclusion, although the environmental marketplace is expected to see strong growth within the United States and other regions of the world, some segments are expected to expand much more rapidly than others. Policy makers would like to see progress in these markets become more technology-driven than regulation-driven, with greater competition in the marketplace. This would encourage innovation and lead to more cost-effective environmental controls, they believe.

Indeed, American construction firms entering this marketplace are finding growing competition from foreign construction and environmental firms, often involving partnerships or other forms of collaboration with other U.S. companies. Because the United States has been a leader in many segments of environmental protection, there appear to be emerging opportunities in other areas of the world where the green movement is catching hold. But capitalizing on these opportunities will require carefully developed strategies, as suggested in Chapter 1.

CHAPTER 6

SHIFTING ATTITUDES CREATE OPPORTUNITIES FOR CONSTRUCTION

SHIFTING ATTITUDES CREATE OPPORTUNITIES FOR CONSTRUCTION

As environmental pressures increase, the attitudes of executives within both construction firms and client companies are shifting. A worldwide survey of the construction industry and its commercial and industrial clients reveals many aspects of these changes. The trends that emerge from this study signal new directions and opportunities for many companies that serve this marketplace.

Traditionally construction firms were only responsible for permitting and satisfying simple environmental requirements at a building site, such as dust suppression, noise control, and the disposal of debris. But the influence of the environmental movement has now become much more invasive. Four different types of needs have emerged for construction companies as a result of expanding environmental requirements.

First is the often urgent need among a wide range of clients to clean up existing polluted facilities. The remediation of contaminated sites has been the focus of a large popular movement, leading to such diverse activities as hazardous waste treatment, underground storage tank testing and remediation, real estate hazardous waste audits, nuclear waste management, toxic waste disposal, and decontamination of contaminated soil and groundwater. While this type of work has only recently begun to emerge as a primary and immediate concern of many industrial construction clients, it has already spawned a rapidly growing, technology-intensive industry.

Second, industry is recognizing the need to retrofit the existing means of production, capturing waste and pollutants right at the production site so as to minimize any release into the air, water, or soil. The captured pollutants or waste often must be treated to make them more suitable for proper disposal. These types of changes generally involve end-of-pipe control systems aimed at providing interim solutions rather than wholesale replacement of production facilities. Customers for such retrofitting projects seek cost-effective solutions since they regard these installations as temporary fixes. Eventually many of these firms will be ready to make a transition to more permanent solutions, creating more of the third type of demand.

The third type of demand is arising as many industrial and commercial clients conclude that entirely new means of production must be developed that can eliminate the need for environmental control systems altogether. This strategy , commonly called "source control," represents the long-term industrial response to the demands of the environmental movement. It calls for owners, engineers, and builders to work together to either eliminate or minimize pollutants and waste, and to either reuse or find alternate markets for as much of the remaining waste as possible.

Finally, industrial and commercial clients have realized that ever-increasing environmental considerations and requirements have begun to constrain site selection, permitting, design, construction, and operation of their facilities. They are increasingly in need of the services of engineering and construction firms with in-depth knowledge in the environmental sector to help them address these concerns and to provide them with new or overhauled facilities without undue expense or unnecessary difficulties and delays during construction or operations. This need is creating the fourth type of market demand.

These four areas of environmentally related needs provide significant new opportunities for construction firms in many different environmental market sectors.

A survey was conducted to help shed light on the reactions of the construction industry and its clients to expanding environmental pressures and to the market potential created as a result. Questionnaires were designed to determine the current involvement of construction firms in the environmental market, identify market areas of interest to them, and probe their views of opportunities and challenges within the market. The survey was also intended to test a number of common perceptions. For example: Is the environmental market currently burdened with excessive legal and regulatory restrictions? Is there a lack of adequate technology? Are environmental services just add-on costs that will eventually be eliminated by market forces (in other words, as added costs, will buyers and sellers ultimately be unwilling to foot the bill for them)? While not everyone subscribes to all of these views, adherents are commonly found throughout society, including the construction industry.

Participants included constructors, designers, technology vendors, and owners of constructed facilities. Constructors included general contractors and subcontractors, while designers included architects and engineers. The questionnaire was sent to a wide variety of firms, large and small, domestic and foreign. Typically they operated in one or more segments of the environmental market, but this did not always hold true (for example, some facility owners did not participate directly in any of the environmental markets). Since a broad survey of the industry was desired, no strict rules were used for selection. Most of the firms appear annually in the *Engineering and News Record* rankings of top contractors, designers, and owners.

Beyond the questionnaires, which played the predominant role in this research, the survey also included several interviews. These, for the most part, confirmed questionnaire results. When results are cited, they include both types of responses. In the discussion of results, the construction industry group includes constructors, designers, and technology vendors. For convenience, these firms are referred to as "construction firms." Since large construction contractors generally have a wide spectrum of capabilities, it becomes impractical to try to sort companies into individual categories. This group, then, represents the supply side of the business, in contrast to the demand side, represented by the owners of constructed facilities.

SURVEY QUESTIONNAIRE DESIGN

Two questionnaires were used in the survey (see Appendices B and C). One was designed for contractors, designers, and technology vendors. The second was tailored for owners of constructed facilities. The main difference between the two questionnaires was the third of three sections, which was modified for the owners. Sections I of the questionnaire asked for general information from the respondent, Section II inquired about the environmental market, and Sections III probed the responses of these firms to the market. Sections II and III each contained several parts. A brief description of each section follows.

Section I: General Information about Your Firm

This section included questions about the age and size of the firm, the role of the firm in the construction industry (e.g., constructor, designer, or owner), and the level of the firm's activities in different segments of the environmental market. This section also inquired about the geographic scope of the firm, including the location of the firm's headquarters and the extent of its operations in various global regions.

Section II: The Environmental Market

Section II contained some twenty questions and was split into three parts. Part A examined the size, location, and type of the total market and its various segments. Specific questions dealt with the dollar size of the total market, market growth rate, relative market sizes in different regions of the world, and future opportunities in various market segments. Part B explored regulatory and legal issues within the environmental marketplace. Particular questions addressed the current level of regulation in the market segments, the degree to which regulations drive the market, and the impacts of these regulations on construction activity. Part C investigated technology issues related to the environmental market. These issues included the role of regulation and government in technology development, the use of proprietary versus off-the-shelf technology, the need for corrective versus preventive technologies, and the degree to which technology represents a barrier to entry in various market segments.

Section III: Response to the Environmental Market

This section consisted of two parts and also contained approximately twenty questions. Although the structure was the same, there were some differences between the constructor/designer and owner questionnaires. Part A examined the business strategies employed by the firms in responding to the environmental market. For the constructor/designers, issues of particular interest were the strategies used to gain the capabilities needed to provide environmental services, the relative importance of the environmental market and its various segments to the firm, and reasons for operating in this market. In the owners' version of the questionnaire, particular questions addressed the importance of various environmental market segments to the business, how normal business activities have been affected recently by environmental regulations and concerns, and the expected need for construction services due specifically to these regulations and concerns. Part B in both versions of the questionnaire looked at the impact of increased environmental awareness and responsibility on the construction industry. Specific questions dealt with the affects of environmental concerns on construction activities, the importance of environmental responsibility for obtaining work in the construction industry, and restrictions placed on construction due to environmental factors.

While the questionnaire contained some 50 questions, it required about 150 answers because many of the questions had multiple parts. Each questionnaire was sent with a cover letter, an instruction page, and a release form for further participation in the survey. Respondents had the option of including or omitting their names and titles. Any material indicating the identity of a respondent was separated from the returned questionnaires, thereby guaranteeing the anonymity of responses.

CHARACTERISTICS OF SURVEY RESPONDENTS

In order to make the results more meaningful, it will be helpful to examine the types of firms surveyed and the characteristics of those that responded. Participants included constructors, design firms, technology vendors, and owners of constructed facilities. Constructors included both general contractors and subcontractors, while designers included both architectural and engineering companies. Questionnaires were sent to a wide variety of companies, large and small, foreign and domestic. Typically, the firms operated in one or more segments of the environmental market, but this did not hold true in all cases. Some owners, for example, did not participate in any environmental market segments.

Three types of lists were compiled. Two of them consisted of contractors, designers, and technology vendors. One included 240 domestic (U.S.) firms and the other consisted of 75 foreign firms. Classifying construction firms as either constructors, designers, or technology vendors can be difficult and somewhat arbitrary. Many firms purport to be all three. By determining the principal roles of the firms chosen, however, the first two lists included some 145 constructors, 145 designers, and 25 technology vendors, for a total of 315 firms. The third list comprised 120 domestic owners of constructed facilities. These owners ranged from hospitals to petrochemical and utility companies.

Questionnaires were sent to the domestic construction firms in August 1991, and the international mailing followed in September. Owner questionnaires were mailed in January 1992. Responses were received from eighty domestic firms, thirteen foreign firms, and twenty-two owners. This corresponded to a 33-percent response rate (80/240) for the domestic construction firms; a 17-percent response rate (13/75) for the foreign construction firms; and an 18-percent response rate (22/120) from the owners. These response rates can be considered as very good, considering the length and nature of the questionnaires. Almost 95 percent of the respondents gave their names and titles, and many presidents and chief executive officers filled out the questionnaires—especially for the domestic construction firms. In almost all cases, the respondents appeared to be well equipped to answer the questions thoughtfully and reliably. Based on the respondents who did give their titles, the questionnaires in many cases were completed by executives directly involved in planning business strategies for their companies. Close to 30 percent of those who responded agreed to further participate in the survey, and this list provided a starting point for the personal interviews conducted following an analysis of the questionnaire responses.

While the questionnaires played a predominant role in the survey, several interviews were also conducted, and these, for the most part, confirmed the questionnaire responses. When results are cited, both types of responses are included. In the discussion of results, the construction group includes the constructors, designers, and technology vendors. For convenience, all three of these are referred to as "construction firms." Even though some respondents fit into one category or another, the larger construction contractors, in particular, have such a wide spectrum of capabilities that it might be misleading to force them into specific categories. This group is referred to in the results as the "supply side" of the business, while the owners of constructed facilities make up the "demand side."

THE CONSTRUCTION INDUSTRY: *THE SUPPLY SIDE*

Most of the participating firms have been active in the construction industry for quite a while, thus representing a very experienced group. Roughly half of the participating domestic firms report annual revenues of $100 million or less. Of the remaining firms, about 30 percent report revenues between $101 and 500 million; 10 percent report revenues between $501 million and $1 billion; and 10 percent report annual revenues exceeding $1 billion. Fifty percent of the

Table 6.1 Classification of Domestic (d) and Foreign (f) Construction Firms Responding to Questionnaire*

Firm Type	Frequency Response (d)	Percentage Total (d)	Frequency Response (f)	Percentage Total (f)
Public owner	6	8	1	8
Private owner	11	14	3	23
Engineering or design firm	54	68	6	46
Technology vendor	19	24	3	23
Supplier	9	11	5	38
Constructor: General	46	58	12	92
Subcontractor	5	7	1	8

*Note that there are 79 domestic and 13 foreign firms responding. Percentages are rounded to the nearest percent.

firms employ 101–1,000 persons, and 44 percent have more than 1,000 employees. The domestic firms represent a wide spectrum of sizes, from relatively small to very large. By contrast, the foreign firms are much larger on average. Seventy percent of the foreign firms report revenues over $500 million (compared to 20 percent of domestic firms) and 77 percent employ more than 1,000 employees (versus 50 percent for domestic firms). In summary, the foreign firms that responded to the questionnaire tend to be rather large and well established, whereas the domestic firms exhibit much broader ranges of age and size.

Table 6.1 shows how seventy-nine domestic and thirteen foreign construction firms classify themselves. The firms checked all of their principal roles, not just a single role; so the percentages shown do not add up to 100 percent. Rather, the percentages given should be interpreted as follows: X percent of the firms that answered this question classify themselves as a *firm type*. For example, 54 of the 79 domestic respondents—or 68 percent—classify themselves as engineering or design firms.

As Table 6.1 shows, the majority of the respondents classify themselves as designers and/or constructors. In addition, almost 25 percent of the respondents classify themselves as technology vendors. Note that some of the firms also classify themselves as owners and suppliers. However, these are typically secondary roles—especially for the domestic firms. The foreign firms are all constructors and tend to be more integrated, probably due to their large relative sizes. Approximately half of the participating firms, both domestic and foreign, report to be subsidiaries of another company (this data is not shown in Table 6.1).

The firms were asked to report the extent of their current involvement in the eight segments of the environmental market. These segments include: hazardous waste, solid waste, sewer/wastewater, water supply, environmental assessment, pollution abatement, energy, and petrochemicals. The firms reported their involvement as a percentage of total business activity. The possible answers were: not at all = 0%, minor = 10%, modest = 30%, major = 31–99%, and exclusive = 100%. The domestic firms display a wide range of activity. For most of the segments, 50–60 percent report minor or no activity, another 30 percent or so report modest activity, and the remaining 10–20 percent report major activity. However, the hazardous waste, petrochemical, and pollution abatement segments show slightly higher percentages reporting major levels of activity. The clear winner is the hazardous waste segment, where over one-third of the respondents report major activity. It is clear, however, that the domestic firms are a well-diversified group. No conspicuous biases appear.

In contrast, the foreign firms show much stronger involvement in traditional construction markets such as petrochemicals, energy, water supply, and sewer/wastewater. The newer market segments—including solid waste, hazardous waste, pollution abatement, and environmental assessment—constitute little or none of their business activity. In this case, the hazardous waste segment is the clear loser. Over half of the foreign firms claim to not participate in the hazardous waste market at all. Of course, this may be a general reflection of a lesser demand for these services in their respective countries. Since the United States has the world's most mature hazardous waste market (a direct result of RCRA, CERCLA, and other comparable acts), this discrepancy between domestic and foreign firms is not surprising. The sample of foreign firms is small, only thirteen in number (four from Japan), so comparing their results to the domestic firms requires caution. But the foreign firms are typically very large. If we compare them with the domestic firms, particularly those of similar size, a distinct difference appears: the domestic firms are much more active in the emerging environmental segments of the past two decades. Figure 6.1 shows the relative business activity of the foreign and domestic firms in four emergent environmental segments.

The domestic firms, on the average, report higher activity levels in hazardous waste, solid waste, pollution abatement, and environmental assessment than their foreign counterparts. Not many foreign firms claim more than a minor involvement in these areas. Proportionately, many more domestic firms report modest, major, or exclusive activity in these four segments.

The participating firms also list the geographical scope of their business. They were asked to indicate the percentage of their work that is local, regional, national, or international. For the domestic firms, the rankings from most to least important are national, regional, local, and international. The national, regional, and local areas represent markets of similar proportions, but the international work lags far behind in importance for most firms. The domestic firms operate predominantly in North America. Other markets are much less important on average. Of the other markets, Latin America, the Far East, and the

THE CONSTRUCTION INDUSTRY: **THE SUPPLY SIDE**

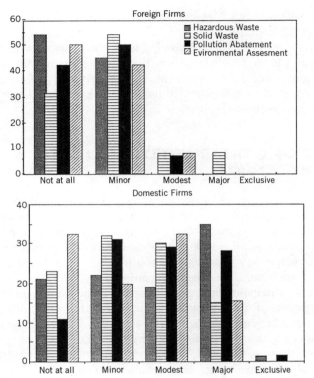

Figure 6.1. Business Activity in Emergent Environmental Segments

Mideast slightly outrank the Eastern and Western European markets in importance.

For the foreign firms, the geographical areas rank similarly except that international markets play a much larger role, in many cases rivaling or surpassing the importance of national markets. The foreign firms typically split work between national and international markets for reasons that seem quite clear. Most of the countries are geographically and economically much smaller than the United States, thus making their own national markets more accessible; but, at the same time, to broaden their total business potential, these foreign firms must rely much more heavily on international markets.

For the domestic constructors and designers, private work appears slightly more important than public work. The foreign constructors and designers rely equally on private and public work. The domestic firms regularly use competitive bidding, cost-plus-fee, and negotiated price contracting, with competitive bidding being the most common. The foreign firms also regularly use all three contracting methods, but competitive bidding and cost-plus-fee play smaller roles than for the domestic firms.

In summary, the domestic and foreign firms that responded to the questionnaire represent a broad range: large, small, diversified, nondiversified, national, and international. The firms operate throughout the world and conduct business in all segments of the environmental market, thus providing a very broad sampling of the construction industry.

OWNERS OF CONSTRUCTED FACILITIES: THE DEMAND SIDE

The owners that responded to the questionnaire are generally large and well-established firms. Over 80 percent of the firms were formed before 1950, and 90 percent employ more than 1,000 persons. The respondents are split evenly between private and public owners. A few respondents reported secondary roles as suppliers, technology vendors, and constructors. Only 14 percent of the respondents are subsidiaries of other companies.

The owners participating in the questionnaire come from many diverse industries. All segments of the environmental market are represented, but the energy segment is the most common: approximately half of the respondents indicate "major" or "exclusive" operation in the energy industry, and the majority of these are electric power utilities. The petrochemical, pollution abatement, and environmental assessment segments are also frequent areas of activity. Some owners are active in the remaining segments (water supply, sewer/wastewater, solid waste, and hazardous waste), but to a lesser extent.

All of the owners are American firms, but almost half of them have operations overseas. Most of the firms operate predominately at the regional and national levels, but the local and international markets are important as well. North American markets are by far the most important. Limited operations extend to Western Europe, Latin America, the Mideast, and the Far East. Eastern Europe is the least popular market for the owners; nearly three-quarters of the respondents report no activity in this region.

PERSPECTIVES ON THE ENVIRONMENTAL MARKET

This section examines the environmental market from the perspectives of both the construction firms and the owners, occasionally focusing on the distinctions between the owners' responses and those of the constructors/designers.

The questionnaire inquired about the environmental market as viewed by the respondents. Three areas of the market were addressed: Market Size, Location, and Type; Regulatory and Legal Issues; and Technology Issues. The results from each of these parts are presented below.

The Construction Industry: *The Supply Side*

MARKET SIZE, LOCATION, AND TYPE

Defining the environmental market is difficult; therefore, any numbers attached to the market must be viewed cautiously. The questionnaire contained a series of questions that asked respondents to estimate the dollar size of their domestic environmental market now and in the year 2000. Similar estimates were requested for the worldwide market at both times. The estimates are reported in Table 6.2. Of more interest than the actual dollar figures are the relative distributions of responses of domestic versus international firms.

It is probably wise not to dwell on specific numbers since assigning meaningful values proves very difficult. The trends indicated by these responses seem more significant. Both the domestic (U.S.) and foreign firms expect significant growth in domestic and worldwide markets, as deduced from the relative shifts in responses to questions concerning current and future markets. Also note that, on the average, foreign firms assess their domestic markets as much smaller than their American counterparts. Seventy-seven percent of the foreign firms estimate their current domestic market to be less than or equal to $10 billion, and, of this total, almost half believe the market to be less than $1 billion. By contrast, only 27 percent of the American firms believe their domestic market to be less than or equal to $10 billion, and none estimate the market below $1

Table 6.2 Domestic and Foreign Responses Regarding Environmental Market Size

"What is the approximate annual size of the environmental market?"	<$1 Billion	$1–10 Billion	$11–50 Billion	$51–100 Billion	>$100 Billion
Domestic Firms					
In your country today?	0	27	31	22	20
In your country in 2000?	0	6	29	36	29
Worldwide today?	0	1	17	29	53
Worldwide in 2000?	0	1	4	18	76
Foreign Firms					
In your country today?	31	46	23	0	0
In your country in 2000?	0	50	34	8	8
Worldwide today?	0	8	25	42	25
Worldwide in 2000?	0	0	8	25	67

*Figures report percentages of total respondents that gave a particular answer. On average, 72 domestic and 12 foreign firms responded to each.

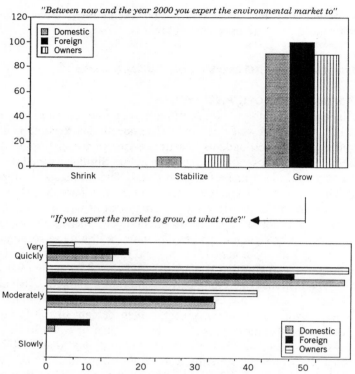

Figure 6.2. Comparison of Views Regarding the Growth of the Environmental Market

billion. Both the foreign and domestic firms agree that the worldwide environmental market should exceed $100 billion by 2000.

A separate question explicitly asked the firms if they expect the market to grow between now and 2000. The respondents overwhelmingly say the market will grow while very few think the market will shrink or stabilize (see Figure 6.2). Of the firms predicting growth, almost all describe the growth to be "moderate" to "very quick," rather than "slow," also shown in Figure 6.2. In fact, close to half of the domestic and foreign respondents predict quick growth (i.e., growth between moderate and very quick).

The respondents classified environmental markets from regions throughout the world using a five-point scale, going from "small" to "medium" to "large." The markets in North America and Western Europe often rank better than medium in size. Expectedly, most American firms list the North American market as large. The Eastern Europe market received mixed reviews, indicating that at least some firms see significant opportunities while others do not. Responses for the Far East are scattered around the medium ranking, while Latin America and the Mideast are generally viewed as smaller markets. It is worthwhile to note that all of these responses more likely reflect actual versus potential markets, but market potential does play a role. The mixed rankings of Eastern Europe provide an example. The potentially large Eastern European market undoubtedly led some firms to label this market as large, despite the fact that most view the current actual market as small.

The respondents indicate potential opportunities for the construction industry over the next five years in each of the environmental segments. For the domestic firms, the solid waste and hazardous waste segments finished at the top. Most respondents believe these segments provide the most opportunity. However, the remaining segments, especially petrochemicals, pollution abatement, and environmental assessment, appear strong as well. No segment was singled out as a clear loser. The foreign respondents view all the segments in much the

Table 6.3 Domestic and Foreign Responses Regarding Construction Opportunities in the Environmental Market

"Indicate the potential opportunity over the next five years for construction in the following areas:"	Percent					
	Minor 1	2	Modest 3	4	Major 5	Don't Know 9
Domestic Firms						
Hazardous waste	3	0	8	32	56	1
Solid waste	1	5	16	42	30	6
Sewer/wastewater	1	9	33	3	16	6
Water supply	1	9	35	36	12	6
Pollution abatement	0	4	27	36	31	3
Environmental assessment	3	16	17	42	17	6
Energy	0	14	33	23	14	16
Petrochemicals	1	11	24	20	20	23
Foreign Firms						
Hazardous waste	0	15	31	46	8	0
Solid waste	0	0	39	39	23	0
Sewer/wastewater	0	0	31	46	23	0
Water supply	0	15	46	39	0	0
Pollution abatement	0	0	31	31	22	8
Environmental assessment	0	8	22	62	0	8
Energy	0	15	15	55	15	0
Petrochemicals	15	31	39	15	0	0

*Figures report percentages of total respondents that gave a particular answer. On average, 78 domestic and 13 foreign firms responded to each part.

same way. In contrast to the American firms, hazardous waste finished in the middle of the pack, and sewer/wastewater and energy are seen to hold the most promise. Once again, no clear losers were identified, but the petrochemical market did receive rather lower rankings than the other segments. Table 6.3 shows the responses to this question.

REGULATORY AND LEGAL ISSUES

The respondents to this questionnaire offered their opinions of current environmental regulation, with somewhat surprising results. Most notable, in all but one of the segments the domestic respondents overwhelmingly view regulation to be "not excessive." Only in the hazardous waste segment was opinion roughly split between the "excessive" and "not excessive" votes. The foreign respondents were more inclined to rank regulations in their countries as excessive, but only one segment—energy—held a majority of excessive votes. From these findings, environmental regulations appear **not** to be excessive for construction firms operating in these areas. The respondents also generally believe that the environmental market should be driven more by regulation forces than market forces.

These results may be explained in part by the nature of environmental effects. Environmental effects, as discussed previously, are largely externalities in most markets. In other words, the costs and benefits to society of environmental effects are not explicitly accounted for in the pricing of most goods and services. Therefore, it should come as no surprise that the respondents to the questionnaire favored regulation over market forces. Generally speaking, many of the environmental market segments such as hazardous waste and air pollution abatement owe their very existence to regulatory forces.

The respondents further believe that regulations should be more incentive-based, not more punitive. The opinions show that while punitive measures should be retained, most favor incentives over punitive measures. They rank the federal government as the most important actor in setting environmental regulations of the future, with state governments not far behind in importance while local governments are seen as the least important player. In addition, most respondents tend to believe that it is important to establish international regulations concerning the environment. In the absence of such international regulations, however, they feel that neighboring countries should have similar regulations. The foreign firms hold these two beliefs with greater conviction, more heavily stressing the importance of international regulations and comparable regulations between neighboring countries. Such a finding seems reasonable since most of the foreign countries are much smaller than the United States and have more neighbors close by.

Environmental regulations may produce additional opportunities for the construction industry, as in the hazardous waste market where regulation has driven

OWNERS OF CONSTRUCTED FACILITIES: THE DEMAND SIDE

Table 6.4 Domestic Firm Responses to Questions Regarding Environmental Regulations

	Percent				
"What is the likelihood that future environmental regulations will result in additional work for the construction industry?"	Low Probability 1	2	Moderate Probability 3	4	High Probability 5
	0	1	13	35	51
	Percent				
"What is the likelihood that future environmental regulations will result in greater restrictions on construction?"	Low Probability 1	2	Moderate Probability 3	4	High Probability 5
	4	16	27	30	23

*Figures report percentages of total respondents that gave a particular answer; 78 domestic firms responded to each question.

site cleanups. Likewise, environmental regulations may inhibit construction activity, as is the case for current restrictions on development in or near wetlands. Of course, some regulations restrict construction in some areas while promoting it in others. The questionnaire did not attempt to judge the appropriateness of specific regulations that either promote or restrict construction. Rather, the respondents were asked whether future environmental regulations would provide additional construction opportunities. A separate question asked whether future environmental regulations would result in greater restrictions on construction. Table 6.4 shows the responses to these questions from the domestic respondents.

The respondents clearly believe that future environmental regulations will both provide additional opportunities for construction and be likely to increase restrictions on construction. The responses to this question reflect general opinion that, while some areas of construction may be further restricted in the future due to environmental regulations, other areas will be opening up as a result. As seen in Table 6.4, higher percentages of respondents expect additional opportunities than expect greater restrictions. Furthermore, there is no significant correlation of answers between the two separate questions: the answers to the two questions are independent of each other. Later questions deal directly with the prospects for given market segments.

Regulations often deal with directly assigning liability for environmental damage. The hazardous waste business offers a good example. Under Superfund legislation, liability provisions cover just about everyone ever involved in the site—including those who sent waste to the site, transported the waste, ran the site, and sometimes even bankers who funded these companies. In fact, the process of establishing liability draws more attention than the actual cleanup of the sites. Under such a system it is not surprising that contractors involved in Superfund cleanups are nervous. If further cleaning of so called "cleaned up" sites is ordered in the future, will the contractors themselves be included in the list of potentially liable firms? This question represents a valid concern. In addition, the insurers of the polluting firms are increasingly being asked to foot the bill for cleanups. Since these costs can be staggering, either companies or their insurance companies could face bankruptcy, or close to it, if they are forced to pay. This threat frequently leads to extensive and lengthy litigation. Establishing liability is fraught with obstacles.

Several questions in the survey addressed the issue of liability as it relates to the environmental market as a whole. Over 80 percent of the domestic respondents believe potential liability is a significant hindrance to operating in the environmental market, with half of this number viewing liability as a "very significant" obstacle. The foreign firms feel less strongly about the issue, but still believe liability is a significant issue. Figure 6.3 shows these opinions concerning potential liability.

The figure indicates that foreign construction firms tend to rate potential liability as a "significant" hindrance, while their domestic counterparts are more inclined to rate liability as "very significant." The lower rating by foreign respondents probably reflects the fact that environmental liability has not been

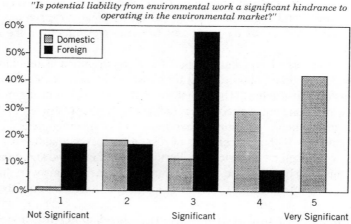

Figures on the vertical axis report percentages of total respondents that give a particular answer on the horizontal axis.

Figure 6.3. Potential Liability as a Hindrance in the Environmental Market

OWNERS OF CONSTRUCTED FACILITIES: THE DEMAND SIDE

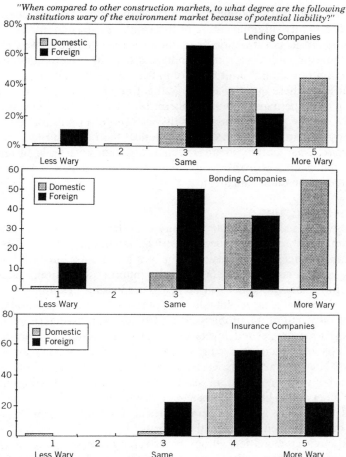

Figures on the vertical axis report percentages of total respondents that give a particular answer on the horizontal axis.

Figure 6.4. The Degree That Lending Companies, Bonding Companies, and Insurers are Wary of the Environmental Market as Compared to Other Construction Markets

as far reaching in most countries as in the United States. However, current debate in Europe, especially in Britain, centers around this volatile issue.

Figure 6.4 clearly shows that domestic construction firms generally believe that lending companies, bonding companies, and particularly insurers are more wary of the environmental market than of other construction markets. The foreign construction firms, while agreeing on the attitude of insurers, generally do not view lending and bonding companies as more wary of the environmental market. The highly litigious nature of the United States may partly explain the wary outlook of its construction firms. In a separate question, the vast majority of both domestic *and* foreign firms indicated that the governments of their respective countries should limit liability in the environmental market.

Responses to these questions indicate that foreign firms see potential liability as a serious, and perhaps growing concern. Moreover, while domestic construction firms already view potential liability as a more serious concern than their foreign counterparts, the difference may reflect the fact that the U.S. environmental market, particularly in areas such as hazardous waste, is currently much more regulated than elsewhere—even in the highly developed countries that participated in the questionnaire.

TECHNOLOGY ISSUES

Technology plays a central role in many of today's environmental problems. As environmental regulations become stricter, technology must evolve to meet these goals as efficiently as possible. The best technology must meet two often-opposing goals: maximum cleanup and minimum cost. Defining either goal is tough; defining both together can be nearly impossible. Defining exactly what is "clean," or what is an acceptable level of risk from environmental pollutants, represents only half of the problem. It is also necessary to determine how much it will cost to meet desired levels of pollution reduction. Unfortunately, the marginal cost of reducing risk soars as the risk shrinks. For some environmental endeavors, such as hazardous waste cleanup or air pollution abatement, a suitable technology may not exist, or, even if it does, it may be prohibitively expensive. For other areas, like solid waste, suitable technology may already be available. But technology advances remain crucial for the successful resolution of many of our present environmental problems.

Should environmental technology be driven by market or regulatory forces? Some argue that when technology is driven by regulatory forces, innovative technologies are slighted. They say that regulation creates an incentive to develop technology that meets only the bare minimum regulatory standards. It is argued that minimal performance is especially encouraged when the technology itself—not just the cleanup standards it must meet—is highly regulated, it is argued. As a consequence, few incentives exist to develop innovative technologies capable of exceeding such standards. Therefore, regulatory forces essentially drive the development of technology. A more effective alternative proposed by those who accept this view is to set cleanup standards and then let the market determine which technology most efficiently meets the needs of customers. The questionnaire results show that the domestic and foreign firms believe strongly that current environmental technology is driven primarily by regulation. If we accept the above argument, the natural conclusion is that innovative technologies in the environmental market are not highly rewarded.

When asked directly whether innovative technologies are generally favored by funding agencies in their countries, two-thirds of both the domestic and foreign respondents answered "no." When these negative respondents were asked how much emphasis *should* be placed on innovative technologies, essentially all (>95 percent) responded that moderate to strong emphasis should be placed on innovative technologies.

The respondents indicate that R&D for environmental technologies is currently ongoing in their countries, with their respective governments playing an important role in funding these efforts. As might be expected, responses indicate that the governments of the foreign countries are playing a more important role on the average. Off-the-shelf technology was cited as being more important than the development of proprietary technology. In addition, the domestic respondents view potential liability as a restriction on the development and use of new technology in the environmental market. The foreign firms, by contrast, did not view potential liability as much of a restriction. This fits the general pattern seen earlier, with liability issues appearing less important to foreign than to domestic firms.

Both domestic and foreign firms share the opinion that, while past technology has mainly been corrective (in other words, developed to remedy previous environmental damage), future technologies will be more preventive in nature—designed to prevent or mitigate harmful environmental effects. The respondents also believe that slightly more emphasis *should* be put on developing preventive technologies, although there is agreement that corrective technologies will remain important. Both groups also indicated strong preferences for government promotion of environmental technology. When asked whether their government should promote environmental technology in their country, over 90 percent of the domestic and foreign firms answered "yes." The domestic firms prefer indirect promotion through incentive programs versus direct government sponsorship. The foreign firms, while also preferring incentive programs, see a greater role for direct sponsorship. Once again, such a difference likely results from country-to-country differences in the government's role in industry. The responses indicated that the vast majority (94 percent) of domestic firms favor government promotion of environmental technology, though the preference is definitely for indirect support through incentives rather than direct government sponsorship.

The respondents were also asked to indicate the competitiveness of environmental technology from the United States, Europe, and Japan. Possible answers included "not competitive," "moderately competitive," "very competitive," and "don't know." Realizing that competitiveness is a relative term and not precisely defined, slight differences in answers are not emphasized here. *Both the domestic and foreign firms give American technology the highest percentage of very competitive rankings.* Both sets of respondents, especially the domestic firms, know the least about Japanese technology. European and Japanese technology receive good reviews among those firms familiar with the technology, but U.S. technology still ranks highest of the three. American environmental technology is certainly more developed in areas such as hazardous waste, where greater restrictions and regulations have demanded more cleanup action than in foreign countries.

Another question asked respondents to describe environmental technology development in their own countries and worldwide on a five-point scale that went from inadequate to adequate. The domestic respondents tend to assess environmental technology development in the United States as more adequate than not, although a few actually go so far as to describe it as "adequate." The domestic firms feel that, on a worldwide basis, environmental technology tends to be inadequately developed. The foreign firms indicate quite strongly that environmental technology in their own countries and worldwide tends toward the inadequate side, though their domestic technology development got slightly higher marks.

Finally, respondents were asked if the cost of environmental technology acts as a barrier to entry in each of the eight market segments. The cost of environmental technology creates the strongest barriers to entry in the hazardous waste segment. The hazardous waste segment received two and a half times more votes for "strong barrier" than any of the other segments. The respondents are least familiar with the cost of environmental technology used in the petrochemical and energy segments. This conclusion is derived from the fact that about a third of the respondents indicated answers of "don't know" when asked if the cost of environmental technology served as a barrier to entry in the segments. Of those familiar with the two segments, most view the cost of technology as a modest barrier to entry. Likewise, the respondents generally view the cost of technology as a modest barrier to entry into pollution abatement. The cost of technology seemed much less of a concern for the water supply, sewer/wastewater, and environmental assessment fields.

OWNERS OF CONSTRUCTED FACILITIES: THE DEMAND SIDE

MARKET SIZE, LOCATION, AND TYPE

In terms of dollar estimates, the owners view of the environmental market differed little from that of the construction firms, as indicated in the previous section. As shown by the responses in Table 6.5, the owners also expect the environmental market to grow. When asked directly, 90 percent of owners

Table 6.5 Owner Responses to Question Regarding Environmental Market Size

"What is the approximate annual size of the environmental market" (Current US Dollars)	<$1 Billion	$1–10 Billion	$11–50 Billion	$51–100 Billion	>$100 Billion
i. "in your country today?"	6%	33%	17%	11%	33%
ii. "in your country in 2000?"	6%	12%	29%	6%	47%
iii. "worldwide today?"	6%	12%	18%	12%	53%
iv. "worldwide in 2000?"	6%	0%	25%	6%	63%

On average, 17 owners responded to each part.

stated that the environmental market will grow between now and the year 2000. The remaining 10 percent picked the market to stabilize, but none expected it to shrink. Of those expecting market growth, all estimated moderate to very quick growth. These views mirror those of the construction firms. A comparison of results appears in Figure 6-4.

Relatively, the largest environmental markets are believed to be North America and western Europe. The other markets listed - eastern Europe, Latin America, the Far East, and the Mideast - are viewed as small to medium sized. The rankings of eastern Europe once again displayed a mixture of opinion, indicating at least a growing interest in this area. These results closely match the findings for the construction firms.

Over the next five years, the owners expect great opportunities for construction in solid waste, hazardous waste, energy, pollution abatement, and environmental assessment. Opportunities in the petrochemical, water supply, and sewer/waste water markets are seen as less certain. In contrast, the construction firms viewed the prospects for the water supply and sewer/waste water segments as strong. The uncertainty expressed by the owners about these segments probably reflects their relative lack of involvement in these areas, rather than a truly questionable future for them. With pending U.S. legislation such as the new Clean Water Act, the direction of both markets could significantly change.

REGULATORY AND LEGAL ISSUES

The owners differ somewhat from the construction firms in their assessment of current environmental regulation. Whereas the construction firms generally believe that regulations are *not* excessive in most market segments (except hazardous waste), the owners view current regulation as more restricting. The owners split votes between "excessive" and "not excessive" choices. Since owners are frequently the target of environmental regulation, it is not surprising to

SHIFTING ATTITUDES CREATE OPPORTUNITIES FOR CONSTRUCTION

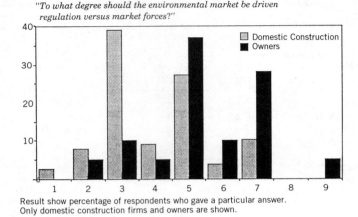

Figure 6.5. What Forces Should Drive the Environmental Market: Market Versus Regulatory

find them more inclined to view regulation as excessive. Regulation directed at hazardous waste is viewed as excessive by most owners, a feeling that also seems to hold for the pollution abatement segment.

The owners and construction firms diverge regarding opinions on whether the market should be driven by market or regulatory forces. Whereas the construction firms seek more emphasis on regulatory forces, the owners definitely prefer an emphasis on market forces. A graphical presentation of these results appears in Figure 6.5.

Figure 6.5 indicates an apparent difference between the domestic construction firms and owners: the construction firms tend to favor regulation while the owners tend to favor market forces. A statistical chi-square test reveals a very high probability (>99 percent, i.e., $p<.01$) that the two groups view this matter differently.

Earlier, we discussed the likelihood that construction firms probably favor regulation because environmental problems are largely externalities. Without such regulation, the market system would not account for social benefits and costs that industry imposes on the environment, and hence little demand for environmental services would arise. Therefore, regulations play a significant role in creating the environmental services market for the construction industry. The externality argument also neatly explains the position of the owners. If no regulations existed, the owners would have to undertake fewer measures—and therefore take on much less cost—to protect the environment. Of course, not all owners view the market this way. And in some cases, a particular firm may save money by voluntarily taking actions to minimize the impact of its activity on the environment—for instance, the firm could save money on disposal costs if it reduced waste generation). As a general observation, however, we see that the results of the questionnaire indicate that owners prefer less regulation, while environmental service firms (here construction firms) prefer more. This is a natural conclusion; regulation typically involves additional cost for the regulated industry, but the same regulation expands the market for environmental service firms.

The owners believed that new regulations should be supported with incentives rather than punitive measures. They saw the federal government as the most important determinant of future environmental regulations, followed by state, and then local governments. The owners also considered international environmental regulations important, but, where they were lacking, owners very much favored comparable regulations across neighboring countries. All of these views correspond to those expressed by the construction firms on the same topics.

Owners indicated that future environmental regulations will very likely lead to additional work for the construction industry. At the same time, they expect future regulations to result in greater restrictions on construction. In fact, the owners felt that there is a greater probability of restrictions than did the construction firms. This is important because owners create demand for construction. They generally believe that greater restrictions are more likely, indicating that demand for construction may be somewhat less than the construction firms themselves expect.

The owners agreed with the construction firms that potential liability from environmental work is a significant hindrance to operating in the environmental market. They also believe that lending companies, bonding companies, and particularly insurers were more wary of the environmental sector than of other construction markets. The owners differed from the construction firms regarding whether the government should provide a broad insurance plan that limits liability for environmental work. Whereas the construction firms strongly favored such a plan, the owners split evenly for and against the idea.

TECHNOLOGY ISSUES

According to the owners surveyed, the development of environmental technology is mainly driven by regulation. These respondents claim that, while R&D of environmental technology is ongoing in their countries (all in the United States for this case), governments play a minor-to-modest role in funding this R&D.

Developing proprietary environmental technology—rather than using off-the-shelf equipment—appears slightly more important to the owners than to the construction firms. This is reasonable since most constructors and designers generally rely on widely available off-the-shelf technology rather than devel-

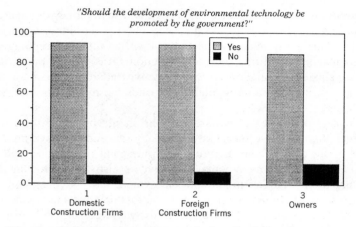

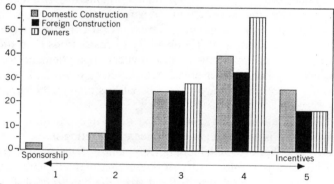

Figure 6.6. Opinions on Government Promotion of Environmental Technology

oping their own. The owners, perhaps with more specific needs and much larger R&D departments, more often favor proprietary technology. They indicate that potential liability has only moderately restricted the development and use of new environmental technology. This view also contrasts with that expressed by construction firms, where potential liability presents a more serious restriction on technology development and use. These different views of potential liability help explain why owners are more likely than construction firms to favor developing proprietary technology.

On the remaining technology issues, the owners generally concur with the construction firms. Their responses indicate that, while past environmental technology was directed more at corrective than preventive purposes, they expect future emphasis to be on preventive measures. They view a proper balance as favoring preventive technology while retaining a good proportion of corrective technology. The vast majority of owners say the government should promote the development of environmental technology, primarily through incentives rather than direct sponsorship. As shown in Figure 6.6, this view mirrors that of the construction firms. Most participating firms favor government promotion of environmental technology, but they generally prefer incentives over direct government sponsorship.

The owners ranked American ahead of European and Japanese environmental technology, but only about half of them were familiar with technology from Europe or Japan. While they generally saw domestic (U.S.) development of environmental technology as less than adequate, they viewed worldwide technology development even less favorably, more towards the inadequate than the adequate end of the spectrum.

The owners indicated that the cost of environmental technology creates at least modest barriers to entry in all market segments. However, they generally believe that these cost barriers are higher for the petrochemical, energy, and particularly the hazardous waste segments.

BUSINESS STRATEGIES

The Construction Industry: *The Supply Side*

Both domestic and foreign respondents indicated the importance of several strategies used to acquire the capability to provide environmental services. A large majority (about 70 percent) emphasized the establishment of in-house capability. Joint ventures also proved important, but to a slightly lesser degree. Joint ventures are particularly attractive when firms want to operate internationally. Joint ventures with local companies often provide better access to knowledge of local regulatory, business, and cultural differences. Mergers and acquisitions play a smaller role for the respondents as a means for providing environmental services.

Environmental work represents a much greater share of the domestic firms' business than it does for the foreign firms. However, the foreign firms expect environmental markets to be more important to them in the next ten years. These findings support earlier conclusions that, on average, the domestic firms have more experience in the environmental market than their foreign counterparts. At the same time, foreign firms are moving to capitalize on the growing worldwide need for environmental services. Figure 6.7 provides some evidence for these claims.

For both the domestic and foreign firms, the primary reason for operating in the environmental arena is the size and growth potential of the market. Approxi-

SHIFTING ATTITUDES CREATE OPPORTUNITIES FOR CONSTRUCTION

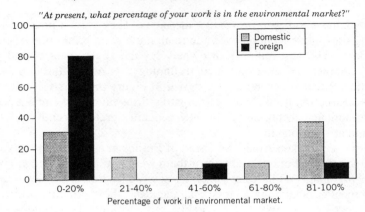

Figure 6.7. Importance of Environmental Work Now and in the Next Ten Years

mately 90 percent of the firms list "market size and growth potential" as a factor that attracted them to environmental work. In addition, the expected profitability from the market attracted many of the firms. The domestic firms appear more inclined to rank expected profitability as a reason to operate in the market, but expected profitability is also important to the foreign firms. Most of the firms that entered the market due to a perceived competitive advantage over their rivals indicated that differentiated service, more than low cost, gives them an edge.

While 70 percent of the domestic firms expanded operations to include environmental services, 100 percent of the foreign firms expanded to include these services. Since the foreign respondents are all very large, well-established companies, this result comes as no surprise. It is interesting to note that a rather high percentage of domestic firms (30 percent) were formed exclusively to provide environmental services. This result probably reflects the generally more mature and specialized environmental markets in the United States, particularly in emergent areas such as hazardous waste.

Domestic firms report that they have grown quickly in the environmental area over the past 5 years, both in terms of employees and revenue. Almost half of them reported growth in "environmentally related" employees *and* revenue in excess of 20 percent over the past 5 years, with many firms (about a fifth) reporting growth over 40 percent. In contrast, half of the foreign firms reported growth in the environmental area between 10 and 30 percent, and no firms reported growth in excess of 40 percent. The higher domestic growth rates are undoubtedly influenced by quickly growing, upstart companies. But the growth rates also reflect a surge in U.S. demand for environmental services over the past 5 years. These very high growth rates will certainly settle down as the markets mature, but they will likely remain strong in the near future.

The questionnaire contained several questions directed at those firms that do not currently operate in the environmental market. Since most of the firms do operate in the market, there were few responses to these questions. However, thirteen domestic and three foreign firms indicated no current activity in the environmental market. Seven of the ten domestic firms had considered entry into the environmental market but then had decided against it. The primary reasons for *not* entering were potential liability, high cost of insurance, and difficulty in obtaining bonding. Furthermore, most of the domestic firms had no current or foreseeable plans to provide environmental services, and only one foreign firm planned to provide such services. Most of the firms use environmental services provided by outside suppliers, and they ranked these services as moderately important for their businesses. While environmental markets are currently attracting more and more construction firms, some (around 15 percent in this survey) have chosen to stay out and express no desire to enter in the future.

Owners of Constructed Facilities: *The Demand Side*

The owners that responded to the questionnaire covered the entire range of environmental work, with each segment represented. Some were active in the petrochemical industry, others in the solid waste business, for example. The most frequently mentioned segments were energy and pollution abatement, with half or more of the respondents claiming to have operated in these segments for more than ten years. Table 6.6 shows the expected importance of the environmental segments for the owners in the next ten years. From the table, we see that energy, hazardous waste, and pollution abatement will be more important than they are now. Environmental assessment and solid waste are also expected to gain in importance. Sewer/wastewater is predicted to be slightly more important, but water supply and petrochemicals are expected to remain about the same. With the recent attention given to hazardous waste cleanup, it is no surprise that firms expect hazardous waste to become even more important in the future. Energy will obviously remain a primary concern of most firms. Moreover, air pollution regulation under the Clean Air Act will certainly contribute to pollution-abatement initiatives.

The owners strongly believe that environmental regulations and concerns have significantly shaped the direction of their industries in recent years. On this issue, all owners answered "moderately" (23 percent), "significantly" (45 percent), or "very significantly" (32 percent); none answered "not at all" or "slightly." This result supports the popular view that industry faced (and still faces) an "environmental revolution" of sorts in recent times. Although the effects varied from one industry to the next, it is doubtful that any industry completely escaped the effects of this increased environmental regulation and concern. The owners generally believe that their production or process activities were affected most by environmental regulation and concern. The firms claim that resource and material input also felt significant effects. The distribution, marketing, and service activities were also affected, but to a lesser degree.

The owners were asked to indicate what facilities, if any, they expect to need *due to* environmental regulations and concerns. Table 6.7 presents the response to this question. As indicated in the table, most owners expect needs for all of these facilities. The question did not address the degree of need for the facilities, but still an important conclusion emerges: Owners expect some need for new

Table 6.6 Importance of Environmental Segments for Owners in the Next Ten Years*

"Compared to the present, how important will these areas of the environmental market be for your firm in the next 10 years?"	Percent				
	Less 1	2	Same 3	4	More 5
Owners					
Hazardous waste	0	0	28	28	44
Solid waste	0	5	26	37	32
Sewer/wastewater	0	11	33	33	22
Water supply	6	6	53	12	24
Pollution abatement	0	0	26	32	42
Environmental assessment	6	0	28	28	39
Energy	0	0	29	18	53
Petrochemicals	24	6	47	12	12

*Figures report percentages of total respondents that gave a particular answer; 17 owners responded to each part.

Table 6.7 Facilities Needed Due to Environmental Regulations and Concerns

"Do you expect a need for new constructed facilities (in your industry) due to environmental regulations and concerns?"	Yes (%)
New or modified production/process facilities	80
Energy supply facilities	85
Air-pollution-abatement facilities	85
Treatment or containment facilities for:	
solid waste	75
wastewater	60
hazardous waste	75

*20 owners responded to one or more parts.

facilities in all these areas *due to environmental regulations and concerns*, which is quite different from normal replacement due to wear and tear.

Essentially all (95 percent) of the owners think recent environmental regulations and awareness has increased the need for environmental services in their industries. Likewise, they expect an increased need for environmental services between now and 2000, with none of the owners expecting a reduced need. In order to secure such services, the owners rely heavily on outside suppliers and on establishing in-house capabilities. Mergers and acquisitions, as well as joint ventures, are much less important for acquiring environmental services. All of the owners (100 percent) use environmental services provided by outside suppliers, and they typically see these services as "important" or "very important" for their business. About 20 percent of the owners considered establishing environmental services in-house but subsequently chose *not* to do so. These owners listed potential liability and the high cost of technology and equipment as the primary reasons for choosing not to provide environmental services themselves.

ENVIRONMENTAL AWARENESS AND RESPONSIBILITY

Environmental awareness and responsibility have increasingly gained worldwide attention in recent years, resulting in much greater concern about human activities and their impact on the environment. The construction industry, which is responsible for altering the environment to suit human purposes, including building any necessary structures and facilities, seems particularly prone to scrutiny for its actions. Construction necessarily involves changing the environment in which we live and work. Will construction firms strive to become green companies? How important is it for these companies to be perceived as friends of the environment? This section considers the implications of increased environmental awareness and responsibility on the construction industry.

The Construction Industry: *The Supply Side*

The domestic and foreign construction firms indicate that, in general, the construction industry is "moderately concerned" with the environmental impact of its activities. However, more firms leaned toward the "very concerned" end of the spectrum versus the "not concerned" end. The industry appears somewhat, though not greatly, concerned about its impact on the environment. The firms also believe the construction industry has received a moderate amount of public scrutiny regarding the environmental impact of its activities. In response, the firms have moderately altered operations. As shown in Figure 6.8 well over 90 percent of the respondents believe the construction industry is likely to be increasingly scrutinized with regard to its impact on the environment. In all of the issues we've discussed concerning the questionnaire, the domestic and foreign firms agree on average, but the responses of the domestic firms do imply a slightly greater impact of these environmental issues. For example, the domestic firms are more likely to claim that the construction industry was "very concerned" about its environmental impact, receives "much scrutiny" of its actions, and has "heavily altered" its practices due to environmental concerns. This slight divergence supports earlier assertions that, broadly speaking, environmental regulations and concerns are more widespread in the United States than elsewhere. However, further analysis indicates that this fact may change.

The domestic and foreign firms agree that, in the future, more public scrutiny and self-scrutiny will be directed at the construction industry regarding the environmental impact of its activities. However, the foreign firms are more inclined

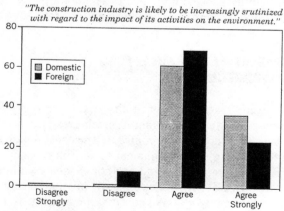

Figure 6.8. Scrutiny of Construction Activities and the Environment

toward "much more" scrutiny. Perhaps this difference reflects a growing concern for the environment in the foreign countries—a concern that previously lagged behind that of the United States.

Both domestic and foreign firms think it is important for the construction industry to be environmentally responsible for the purpose of obtaining work as well as for public relations, with public relations getting the nod as the more important of the two. Domestic firms tend to believe more strongly than foreign respondents that environmental responsibility plays an important role in achieving both objectives. If foreign environmental concern is growing, then foreign firms may find environmental responsibility equally important in the near future.

The firms were asked to indicate their level of agreement with the statement, *"Engineering, design, and construction firms should actively try to promote themselves as stewards and protectors of the environment."* The respondents overwhelmingly agreed with this statement, but 62 percent of the foreign firms "strongly agreed," compared to only 36 percent of the domestic firms.

The questionnaire also asked whether construction firms will be able to create a significant market for themselves as green firms. In other words, will those firms that demonstrate a commitment to environmental awareness and responsibility realize a competitive advantage over seemingly less environmentally concerned firms? Two-thirds (66 percent) of the domestic firms responded "yes," but 91 percent of the foreign firms answered affirmatively. A statistical chi-square test indicates a greater than 95-percent chance ($p<0.05$) that a difference in opinion exists between these groups. Thus, it is highly likely that foreign firms, when compared to domestic firms, generally expect to gain a greater advantage by portraying themselves as green companies. Even more interesting are the owners' responses. Figure 6.9 shows the comprehensive results of this question. Of those construction firms answering affirmatively, most believe this green image will be moderately to very important for public *and* private construction contracts.

The domestic and foreign firms agree that new construction projects have been "moderately restricted" in recent years due to environmental concerns. No firms answered "not restricted" to this question, and only a few (15 percent) believe projects have been "very restricted." Figure 6.10 shows that approximately three-quarters of the foreign and domestic respondents think future construction projects will be "more restricted," while the remaining respondents believe such restrictions will be the "same." None of the respondents think projects will be "less restricted." Moreover, those firms expecting additional restrictions believe there will be *quite* a few more.

Finally, the firms were to indicate their level of agreement with the statements: *"Construction companies should decline to bid or undertake a project which—according to widespread public and scientific opinion—is believed to* be detrimental to the environment," and "The construction industry will not suffer as a consequence of increased public pressure due to environmental concerns of its actions." The domestic firms were fairly well split on these two questions. Approximately 60 percent agree with the first statement and 64 percent disagree with the second. On the other hand, 75 percent of the foreign firms agree with the first, but almost 75 percent agree with the seconds as well. From these responses, it is clear that some firms believe projects should be avoided due to their environmental consequences. However, saying so and doing so are quite different. It is not clear if these companies would, in fact, decline to bid on such a project if they really needed the work. The first state-

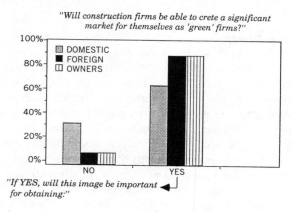

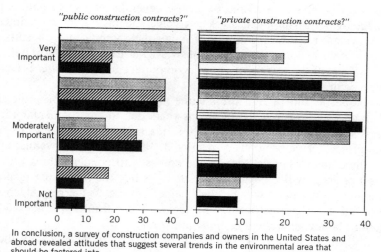

In conclusion, a survey of construction companies and owners in the United States and abroad revealed attitudes that suggest several trends in the environmental area that should be factored into.

Figure 6.9. Competitive Advantage of "Green" Constructtion Firms

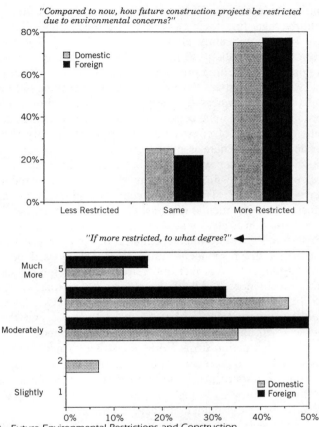

Figure 6.10. Future Environmental Restrictions and Construction

ment, more than most, probably evoked "politically correct" responses rather than true opinions. However, there certainly appears to be some level of agreement with the statement, even though the percentages reported may be artificially high. Results from the second statement show different views between the foreign and domestic firms. The domestic firms marginally favor the belief that the construction industry will suffer due to increased public pressure regarding the environmental impact of its activities; the foreign firms apparently do not agree.

Owners of Constructed Facilities: The Demand Side

The owners indicate that, in general, owners of constructed facilities are "moderately concerned" to "very concerned" about the impact of their construction activities on the environment. The firms also think owners have received quite a lot of scrutiny regarding their construction projects, and that construction projects have been significantly altered due to environmental concerns. The vast majority of owners (about 70 percent) view environmental responsibility as "very important" for selling products and services and for public relations. Quite naturally, the firms stress the importance of environmental responsibility for public relations. The owners definitely tend to rank environmental responsibility as more important than the construction firms. Such a result seems plausible. After all, the owners of constructed facilities are directly responsible for construction decisions. If there is public opposition to a proposed project (or business activity) the owner will hear about it first and must face the direct consequences of such opposition, including the possibility of reduced demand for its products or services. Therefore, owners are more acutely aware of the potential effects of environmental responsibility on their business.

Like the construction firms, owners expect more public and self-scrutiny in the future regarding the environmental impacts of their construction activities. Essentially none of the respondents think this scrutiny will be "less," the "same," or a "little more." In fact, most firms expect "more" scrutiny while some expect "much more." All of the owners agree that "owners of constructed facilities should actively try to promote themselves as stewards and protectors of the environment." Half of them "agree" and the other half "strongly agree" with this statement. Once again, this stands in contrast to the responses of the construction firms. Most construction firms agreed that they should act as stewards and protectors of the environment, but a significant number disagreed (about 20 percent). In addition, a smaller proportion of construction firms "strongly agree" (36 percent). Here again, we see that owners are more sensitive to their environmental image than construction firms. This fact likely reflects the owners' greater reliance on such an image to sell their product or service. Put more properly, the owners think that a *negative* environmental image would have damaging consequences on the sale of their products and services.

The owners were asked if construction firms would be able to create a significant market for themselves as green firms, and an overwhelming 91 percent answered "yes." This response stands in contrast to that of the domestic construction firms, but matches that of the foreign companies. A chi-square test indicates a very high probability (about 98 percent, or $p<0.02$) that a difference of opinion exists between the owners and the domestic construction firms. Essentially all of the owners are domestic firms, so this comparison pits domestic supply against domestic demand. Figure 6.9 shows the results of this question for the domestic and foreign construction firms and the owners.

This difference between domestic construction firms and owners represents an important finding. Owners, who drive the demand for construction, feel more strongly than construction firms that greater market opportunities exist for construction firms with green images. While a significant portion (66 percent) of domestic construction firms agree, proportionately more owners and foreign construction firms believe this claim. This finding strongly suggests that environmentally responsible construction firms could create a competitive advantage over other firms. In a subsequent question, an overwhelming majority (95 percent) of the owners said that they would be more likely to hire a construction firm that has an established record of commitment to the environment versus ones that does not (the firms were presumed to be otherwise comparable).

The owners feel that in recent years construction projects have been somewhat restricted due to environmental concerns. Their answers ranged from "moderately restricted" to "very restricted," with each receiving about the same proportion. They also believe some of the firms expect more restrictions. None of the owners expect fewer restrictions on future construction.

In conclusion, a survey of construction companies and owners in the United States and abroad revealed attitudes that suggest several trends in the environmental area that should be factored into strategic planning for the future. Market growth is expected to be rapid in the United States and in other countries, with every segment expected to expand, with major business opportunities in sectors such as hazardous waste and solid waste in the United States. Surprisingly, most respondents do not see environmental regulation as excessive, although the owners, who are responsible for their facilities, tend to feel they are more excessive than construction firms think they are. The respondents are almost unanimous in expecting to grow further in the future, although most would prefer to see greater use of incentives rather than punitive regulations. They generally agree that the punitive approach suppresses innovation and that innovation is needed for the future. They see a shift coming from the corrective technologies that dominated in the past toward more preventive technology, but indicate that innovation in both areas is important. The owners feel more emphasis is needed on off-the-shelf equipment, while the construction companies gave slightly more importance to proprietary technology. U.S. environmental technology ranks high with firms around the world in comparison to technology for other areas, such as Japan and European countries, especially in the hazardous waste market segment.

Owners agree that environmental regulations and concerns have shaped their industries in recent years, and a large number of them indicate that they expect some need for new facilities simply due to environmental requirements. The great majority of owners also indicate that there would be significant market opportunities in the future for green construction firms with a reputation for environmental commitment, especially in foreign markets. Construction firms did not rate this potential as highly, suggesting that many may not yet recognize this emerging market opportunity.

CHAPTER 7

CORPORATE STRATEGIES AND FUTURE TRENDS

CORPORATE STRATEGIES AND FUTURE TRENDS

The previous chapters demonstrate that, while the environmental movement has led to constraints on the construction industry, it has at the same time created many new opportunities. Construction firms can offer valuable assets for this emerging marketplace, even though the dynamics are quite different from their traditional businesses. Operating in the environmental arena requires considerable adaptation. The capacity to adapt can, in itself, become a powerful tool. This includes learning to work with government agencies at the federal, state, and local levels, gaining insight into policy and regulatory directions, and adding the new specialized skills and knowledge pertinent to any particular environmental sectors chosen for entry.

Some companies are penetrating these new markets by combining existing strengths, such as site planning, earth moving, equipment installation, and project management, with more specialized expertise. These additional capabilities within chosen sectors of the environmental market may be acquired in a variety of ways, such as through acquisitions, collaborations or partnerships with other firms, or by forming new divisions. Many construction companies are already capitalizing on the new market potential by using approaches such as these.

But executives in other firms, noting the complexities, risks, and liabilities involved in environmental work, have avoided entering this tough marketplace. They ask: "Why should our firm get involved?" Even without plans for direct entry, there is one clear reason why every construction company should become familiar with the environmental sector. Firms that thrive in the construction business immensely value the interests of their clients, because they know that, by looking after these interests, they are likely to be rewarded with subsequent contracts not only from satisfied clients but also from others who learn of their performance. This alone makes it essential to become familiar with the environmental area, because it is clear that regulatory requirements will increasingly intrude into construction projects of all types. In addition, more and more major industrial customers will be seeking to build and equip plants and facilities in ways that can help them reduce or even eliminate waste and pollution. They will be looking for contractors that can offer them advice and expertise, starting from initial planning and going through all phases of construction right on into maintenance and operations. Even if a project does not directly involve environmental considerations, clients will expect constructors to be knowledgeable in this area. If clients are unaware of expanding regulatory requirements, they will appreciate helpful advice from contractors, not only to help them meet present and future regulations but also to find ways to minimize the costs of managing waste. Those construction firms that make an effort to meet this obligation to their clients may gain a strong edge on competitors that lag in the environmental area.

Strictly from a business and competitive standpoint, construction firms must gain enough expertise so that they can at least provide useful consulting about environmental requirements to clients seeking solutions. But it is not enough simply to become familiar with the current regulatory status. It is vital to continually keep on top of environmental trends because of the transitory nature of this marketplace. Regulations change frequently. Their scope is also steadily expanding to cover an ever-widening span of industrial and other activities. This means that investments made in response to existing regulations may become obsolete or inadequate well before an owner's capital is amortized. Technologies are also changing rapidly, not only for cleanup and treatment but also for manufacturing processes that can limit or eliminate pollution and waste. A client who invests in technology that quickly becomes outmoded may be forced to make additional investments to upgrade facilities in order to remain competitive. The blame for such short-sighted errors may well be aimed at construction contractors, because clients expect them to be aware of trends in regulations and/or technology.

For these sound business reasons, then, construction companies, must find out about environmental regulations applicable to their own operations and to those of their clients, and they must keep abreast of developments that will drive future policies and regulations. Making intelligent recommendations requires foresight based on familiarity not just with regulations but with a variety of trends: in public attitudes, grassroots movements, global accords, and technology. The shift from an emphasis on command-and-control regulation toward greater use of market-based incentives is likely to accelerate changes that are already reshaping the environmental sector. Industry is looking beyond end-of-pipe solutions toward new methods for source control. Regulation is becoming more integrated, so that past practices involving transferring wastes from one form to another, such as burning contaminated solids and thus releasing toxic fumes, is no longer being permitted in many cases. Future regulations are likely to increasingly push industry toward waste reduction, probably through a combination of penalties and incentives. This may require refurbishing of existing plants and facilities, and the use of new process technology and methods in future plants.

Understanding developments and trends such as these will become more and more essential if a construction firm hopes to provide effective service. This is not to diminish the importance of existing skills and experience. Although innovation will increase in importance, it is not always easy to put new ideas to work successfully. Solutions often depend on the effective coupling of existing capabilities with new technology and methods. To those firms willing to take the risks and adapt, the growth potential may be worth the effort, both domestically and abroad.

DEALING WITH RISK, AND LIABILITIES

Because the nature of risks and liabilities differ in the environmental field from those in traditional construction markets, a firm deciding on entry into any segment of the environmental marketplace must carefully assess the risks and liabilities and develop strategies for managing them. This is particularly important for operating in the fast-growing Superfund cleanup area. It is important to note that potentially responsible parties (PRPs) extend well beyond those who actually created the hazardous or toxic pollution in the first place. PRPs may include the owners and operators of hazardous waste sites (past and present), transportation firms that moved the waste, and even financial institutions that may become responsible for a property through foreclosures. Most important to the construction industry is that cleanup contractors can also be held liable in some cases. This could be potentially devastating, particularly to larger firms entering the market, because legal attacks are often aimed at the party or parties with the deepest pockets even though major responsibilities for the contamination may lie elsewhere.

The forms of potential PRP violations covered under CERCLA are based on the principles of retroactive, strict, and joint and several liabilities. Retroactive liability means that even if a party complied with all existing regulations at the time of the waste disposal, that party can still be held responsible for subsequent damages and the cost of a cleanup. Strict liability means that a party can be held accountable for any damage or safety problems even if every reasonable precaution was taken in doing its part of the work. Negligence does not have to be proven. Further, joint and several liability means that any one of the PRPs, no matter what role was played, can be held responsible for the full costs of cleanup. The availability of surety bonds for cleaning up hazardous and toxic waste sites has almost dried up as a result of joint and several liability. This provision in the law makes it too difficult for a surety underwriter to judge the extent of the risk in case of default by the insured contractor. Some types of injury, such as increases in the incidence of cancer, may be latent for decades before the scope of the injury becomes apparent. Even if the injury is detected early, judgment on the scope of the injury and the size of any award to plaintiffs is a vague area that is decided by the courts. There is no time limit on when these claims may arise or on when government agencies may decide that some party or parties are at fault and fines can be assessed. Not only could contractors and subcontractors be exposed to unlimited liability, but the work might have to done in the face of great uncertainties about the effectiveness of available treatment technologies. Insurers have naturally been reluctant to take on such poorly defined risks.

In an attempt to deal with these issues, Public Law PL 101-584 was enacted in November 1990 to amend CERCLA and provide some relief for sureties of contractors working at cleanup sites or removing contaminated material. This amendment does remove some of the risk, limiting liability in case of a default to the excess costs required to complete the project according to plans and specifications, for example. The surety provider's liability is also limited to only the insured contractor, and the surety provider is protected from third-party claims. But no time limits were set for completion of performance bonds or for initiating suits. Another potential problem is a provision that was included to avoid preempting state laws; therefore, contractors and their insurers must be aware of applicable state laws when assessing the risks involved in a particular project. Another provision limits the protections offered to sureties in the amendment, indicating that they will only apply to a surety acting as a surety. Thus, if the surety agrees to a takeover of the project by an alternate contractor and toxic materials are released during ensuing work, the surety could be sued for unlimited damages because of its action in changing contractors. This means that, in case of a potential default on proper completion of the work, the surety would most likely allow the government to step in and finish the project so that it would have to pay only the excess costs. Because of such remaining limitations on the exposure of underwriters, surety bonds remain virtually impossible to obtain for hazardous and toxic waste work. Further legislative and regulatory changes may be needed to attract underwriters because of the pressing need to find contractors to do this work, particularly at a growing number of DOD sites.

The current system also dictates strongly against trying any innovative methods of treatment because of the increased risk involved. There is likely to be some rationalization of the whole system of liability assessment, however, because of the government's difficulties in finding qualified contractors for decontamination and detoxification projects. The larger construction and environmental firms, whose project-management skills may be needed along with their technical knowledge and equipment, are the most at risk under the present extended liability rules. Even when cleanup costs are assessed to the original companies that generated the waste, these firms feel that the process is unfair because they may have followed all applicable regulations at the time the waste was generated. As indicated in Chapter 4, the EPA has been working to develop new approaches that might define and limit potential liabilities for contractors or subcontractors that take on specific portions of cleanup projects. But in the interim, what can a construction firm do to limit its own risks and liabilities in environmental projects?

The strategy must encompass a wide assortment of potential risks. Litigation between various parties to environmental remediation projects has become common, for example. Suits might be between owners and contractors, site workers and employers, and cleanup contractors and designers or engineering firms.

These suits could involve breaches of contract, project delays, or failures to meet specified performance standards. Legal actions such as these, and fear of them, add greatly to the costs of operating in this arena and often lead to extensive delays during the course of a project. Contractors and other parties involved are also subject to penalties or lawsuits instituted by the EPA or other federal or state agencies under a growing array of environmental laws. More money is frequently spent on legal wrangling than on the actual cleanup work.

There are several ways to structure entry into this field so as to minimize the potential for such litigation and penalties, however. One of these is the design-build, or turnkey approach now coming into favor with some contracting agencies. The agency may seek to hire a contractor that will do the planning and design work for the project and then will also be responsible for managing the actual work on the site. An alternative to this approach, since a single contractor may not have the skills and equipment needed for all phases of a large project, is the partnership approach. Firms with complementary capabilities can form a team to carry out various aspects of the project in collaboration with each other. The owner can also become part of an alliance that might include an engineering/design firm, an environmental specialist, and a construction company. During formation of the alliance and while preparing for the work, it is important for the participants to develop trust in the other members. The team can then work together toward shared goals, with each member sharing in properly assigned risks that are agreed upon among the partners. In such cases, it is important that at least one member of the team have pertinent experience and knowledge in the environmental field, particularly in areas of concern to the specific project. This partner can help other members not just with its technical expertise, but also in areas such as safety and health, and in meeting requirements of applicable federal, state, and local statutes, ordinances, and codes. Workers at a hazardous waste site should at least have completed OSHA hazardous waste health and safety training and any required refresher courses, for example. By working as partners, many of the delays resulting from conflicts among various parties can be avoided and work can be completed on much tighter schedules. It is important to assign clearly areas of responsibility early in the project and to set up mechanisms for continuous interface between those who must work together on some portions of the work. Budget and schedule reviews are also important as work progresses, and each member of the team should be encouraged to contribute useful ideas for improving project methods and to come up with cost-saving measures.

The partnership approach might be applied to a wide range of environmental projects, including water or wastewater treatment systems, process modifications to achieve waste minimization, cleanups of hazardous or toxic waste sites, management of underground storage tanks, or solid waste management projects (such as landfills with impervious linings, recycling facilities and waste transfer stations, or resource recovery plants).

Partnerships between owners and contractors, particularly for massive projects or for cases in which multiple sites with similar needs are involved, might prove particularly advantageous. The partners could share in funding the development of necessary technology, and then contracts for managing various phases of the work, or for work at multiple locations, might result. If contaminated sites are involved, the owners must work cooperatively on the allocation of potential liabilities within the contract. Contractors must be prepared to reject any arrangements that expose them to excessive risks.

Some considerations of contractors, owners, and regulators when applying new technology in the hazardous waste remediation area are suggested by Peter D. Arrowsmith, president of Brown & Root Environmental. Contractors must understand the main technologies available for possible remediation of a site and should be able to delineate the technical and cost risks and rewards of each alternate approach to the owner. They also need to have enough knowledge of the regulatory requirements and the potential health and safety risks to convince state and federal regulators of the effectiveness of the chosen remedial technology. The contractor also needs to assemble the necessary project management, contracting, worker health and safety skills, and quality assurance capabilities to do the work safely and efficiently. At the same time, the owner must be willing to assume ultimate responsibility for the site conditions existing before the work begins.

In the future, these construction firms are likely to become involved in many more projects involving the reduction of pollution at the source. In some cases, this may involve a complete redesign of a process to eliminate sources of pollution. An example of this is the Mobil-Badger process for making ethylbenzene that substitutes a zeolite for the traditional aluminum chloride catalyst, thus eliminating a difficult-to-treat waste stream. Minimization of waste and pollution can also be accomplished by installing more efficient controls or increasing the maintenance schedule for equipment. Sometimes hard-to-treat pollutants can be removed from the waste stream in a pretreatment stage before it enters the final treatment stage. By-product recovery will also be increasingly used, such as saving the oily sludges that collect in a refinery for use in coke products. Recovery for the purpose of recycling or reuse will also become much more common. For example, spent lubricants can be purified and reconstituted for reuse in some cases.

A construction firm that begins to assemble capabilities in any of these areas may find an expanding market of eager customers as environmental pressures increase.

STRATEGIC VISION NEEDED

Even with strong technical and management capabilities, success will still depend on strategic vision. Existing strengths of a firm can be enhanced by acquiring the specific skills, technology, and experience needed to operate within chosen sectors of the environmental marketplace. While business acumen will be important, it will not be sufficient, however, because every sector is in flux. Once a strategic plan is developed, it will be essential to constantly monitor overall developments and trends pertinent to chosen market sectors. Changes in the scope and type of regulations may have wide influence well beyond the direct impact on areas such as waste disposal and treatment. Shifts may also reverberate to auxiliary matters such as insurance, financing, and bonding requirements.

Even so fundamental a notion as how to value long-term investments is being challenged as the concepts behind sustainable development take shape. Sustainable development extends the time period for evaluating investments from the five to ten years typical in today's business world out to generations. To look toward the future, it is helpful to examine the debate now going on over how environmental considerations can better be factored into the economic system, making them a more integral part of normal business decisions. Future directions for legislation, regulatory changes, and incentives to improve environmental performance are likely to emerge from such debates.

Economists still have not yet settled on a uniformly accepted interpretation of what sustainable development actually requires. Although there is agreement that it is a responsibility of the present generation not to take actions that will curtail the quality of life for future generations, there are widely differing views on where the limits should be set. A central controversy pits "neoclassical" economists against what might be called the "ecological" school. The neoclassical view is that future generations should have at least as much capital wealth as exists at present, while the ecological view is that the present generation must ensure an equal share of not just man-made capital but also existing natural assets.

Under the former view, there is no obligation to preserve any particular physical or natural resources, such as forests or fossil fuels, as long as other assets, such as infrastructure and education, are enhanced. One argument supporting this view is that there is uncertainty about which resources will actually be preferable to future generations. As an illustration, in the nineteenth century, some countries left minerals such as coal in the ground for the benefit of future citizens. Because technologies changed, however, nations whose economic strength depended on resources such as these declined while much poorer countries that were found to have large oil reserves achieved unprecedented wealth. Even resource-poor nations, such as Japan, have achieved great economic success through education and training and effective organization of their workforces and workplaces.

In the ecological view, by contrast, it doesn't make sense—and is perhaps even immoral—to compensate for environmental degradation and resource depletion by substituting man-made wealth as the legacy to be left to future generations. Several arguments have been presented to show why such a substitution is not a feasible approach to sustainable development. Some natural assets, like the ozone layer or rain forests, cannot be replaced and are thus quite distinct from the normal factors of production. In addition, they may provide multiple functions. In the case of rain forests, for example, these include climate regulation, watershed protection, and biological diversity. There is also uncertainty about the value of natural assets in the future and about their role in supporting life. It is argued that these assets should be preserved to avoid the potential risks involved in losing them until their functions are better understood. Unlike man-made assets, the loss of natural assets is often irreversible: once they are gone, they cannot be replaced. Preservation may also be vital to the world's poor, especially in developing countries, because they are often heavily dependent on natural assets and their livelihood may depend on them. Another factor that needs to be considered is resilience. Natural systems are more capable of adapting to stress and shocks if there is greater diversity, so it is important to retain diverse species and ecosystems.

Even if agreement could be reached on preserving natural assets, finding practical ways to protect them is still a difficult problem. Some economists argue that monetary values must be assigned, even though they may be hard to determine, so that environmental costs and pricing for various options can be factored into economic decisions. Others feel that trying to assign monetary values to amenities such as wilderness areas, beautiful scenery, and clean air and water is an illusory goal. Preservation will depend, instead, this group feels, on effectively curbing destructive practices and enforcing conservation measures rather than struggling to force fit artificial valuations into an unreceptive economic system.

PRESERVING AND EXPANDING WEALTH

It is clear that current economic rewards and penalties—and the traditional methods of accounting for them—is biased against environmental preservation. The macroeconomic approach to measuring national wealth is a good illustration of this. Income earned from using up natural assets is entered into GNP calculations, but no account is taken of any consequent loss of productive capacity. When timber is cut and sold, for example, GNP is credited for the sale but there is no corresponding debit for the deterioration of the forest. Sustainability requires that some portion of the income derived from the yield on capital stock be invested to replace depleted resources, analogous to the use of depreciation to build up capital funds for replacing machinery as it wears out. But GNP accounting fails to encourage such resource preservation. In fact, many negative consequences of pollution and waste are actually considered to be additions to the national wealth under present accounting methods. While expenditures to combat pollution are counted as income, the negative impact of pollution is not even considered. In 1989, about $2 billion was added to the GNP as a result of the Alaskan oil-spill cleanup effort, for example, even though this may have been the worst environmental disaster in U.S. history! Similarly, the health costs induced by air pollution are added to the GNP.

There have been proposals to adjust GNP (or GDP) calculations so as to evaluate only "sustainable" income, or net national product. Deductions would be made from national accounts for measures to combat pollution and for the value of any pollution damage that could not be mitigated. Income would also be reduced for depreciation of man-made and environmental capital. Such an accounting approach would penalize nations that increased current income at the cost of excessive pollution and of running down capital stock. Already nations such as Norway, France, and Japan include adjustments for natural-resource accounting in their national accounts. Some argue that factors measuring the quality of life should also be included in accounting for national and intergenerational wealth. One example of such a measure is the United Nation's Human Development Index (HDI), a rough indicator that includes longevity, knowledge, and the populace's command of resources needed for a decent life. Some countries with a high GNP (such as South Africa) score much lower in comparison with other nations in HDI, while others with a low GNP (such as Costa Rica) score much higher in HDI. "Development" under an economic system that considers natural and human values would go beyond simply measuring the growth of man-made capital and assets to include all types of enhancements to and depletions from national assets of all types.

PLACING VALUES ON THE ENVIRONMENT

While some efforts are being made to track environmental performance at the national level to encourage sustainable development, it may prove much harder to influence private investments. Decisions on new projects depend on financial analyses that match costs against expected returns on the investment. Typically an investment must be made immediately while returns are expected over some future time period. Since a dollar today is considered to be worth more than the promise of a dollar some time in the future, the value of the projected returns are reduced using an assumed discount rate. For a simple investment, such as for a new machine, the interest rate a bank would charge to borrow the capital is commonly used as the discount rate. The results of such discounting can be dramatic. Today's value of the potential return of a dollar 20 years from now, assuming a discount rate of 10 percent, would be only 15 cents. Some economists insist that environmentally sound investment decisions will be made only when appropriate monetary values are assigned to natural assets (or negative liabilities, such as waste or pollution) so that they can be factored into such analyses.

There is further debate among economists as to whether a higher or lower discount rate would be more appropriate for such purposes. A high discount rate would mean that a future generation would receive less of the benefits and bear a higher share of the costs than the generation making the investment. At the same time, however, a higher discount rate would serve to reduce the number of economically viable projects so that fewer natural resources would be depleted. Some environmentalists have argued instead for a low discount rate, or even a zero rate, so that a dollar in the future would be considered just as valuable as a dollar today. Others argue that there is no effective way to assign monetary values to environmental resources, and that other, nonfinancial methods will be required to protect them for future generations.

Various methods have been proposed for assigning environmental values, each of them with some limitations. If an environmental factor could be subjected to trading in an open marketplace, then theoretically the transactions that took place would eventually result in correct pricing versus other goods. In some cases, practical reality might approximate such a theoretical model if political and other pressures were tempered so that competitive markets could develop. Disposal costs for solid wastes, for example, would be likely to be determined by the marketplace if there were numerous available waste sites competing for the same business. Even in this case, however, without government waste-disposal regulations and enforcement, there might be widespread "midnight dumping" to avoid paying any fees at all. In some cases there may be little awareness of environmental values, and other amenities might have value only to future generations. Some suggest that property rights might be assigned to encourage competitive bargaining, such as for controlling pollution. But how could property rights be assigned to the ozone in the ozone layer, for example? Thus, although there may be limited possibilities for creating an open marketplace and achieving rational pricing in some environmental sectors, the prospects certainly do not appear to be universal. Even in those cases where it does appear feasible to develop market mechanisms, some government regulation or involvement may be required.

FROM REGULATIONS TO INCENTIVES

A brief review of some of the main types of regulation, and of the most commonly suggested incentives, reveals the difficulties faced by policy makers hoping to encourage sustainable development in the future. Command-and-control regulations were the initial approach used to get immediate reductions in pollution, but now there is a transition taking place toward various types of incentives. In a regulatory regime, government can set performance standards, either limiting the total pollution allowed or requiring each firm to reduce its pollution by a certain amount, or it can set specification standards, which dictate the technology that must be used for pollution abatement.

Regulations have been criticized on the basis of a lack of equity, efficiency, and incentives for improved performance. If percentage reductions are required of all polluters, then those firms that have already made improvements would be penalized. If allowable levels are set based on company size, small and growing firms may be at a disadvantage because of greater difficulty in obtaining needed capital. If the total amount of pollution is restricted, then, once the allowable level is reached, newer plants with more efficient control technologies might be prohibited from entering the marketplace. If all firms are required to reduce pollution by a certain amount, then the costs of abatement will be unequal because some may be much better at cutting pollution than others. Further, once required levels have been met, there is no inducement to lower emissions any further, thus stifling the quest for improved methods and new technology. Specification standards also stifle innovation by dictating the use of existing control technologies. A more efficient system would allow each polluter to seek out control equipment or prevention methods that provided the maximum abatement at the minimum cost. If policy makers, recognizing these limitations, tried to fine-tune the rules to make them fairer and more efficient, the resulting complexity would most likely make the regulatory job unmanageable.

Although there are some areas where regulation is probably the only effective solution, in recent times there have been attempts to build incentives or penalties into the economic system for environmental improvements wherever it might be feasible. The types of incentives most commonly suggested include subsidies, emissions taxes, and emissions permits. Each of these also has its strengths and weaknesses.

Firms could be rewarded for acting in environmentally beneficial ways through the use of subsidies. One approach might be to give tax credits for installing waste-treatment or pollution-control equipment. Another would be to provide payments based on the amount of pollution abated. Resistance is strong to using this approach to environmental improvement, however. While subsidies are often used to support policy goals, many feel that tax money should not be used to pay polluters. In a sense, society would actually be promoting environmentally damaging products by assuming some of the social costs of the goods. If tax abatements were offered, it is not clear that the money would be used in the most efficient way to reduce pollution. It would also be difficult to determine just how much pollution was being abated. If payments were related to the amount of pollution reduction, they would resemble farm subsidies for not growing certain crops, and this has not proven to be a very cost-effective way to solve overproduction problems. Companies should not be encouraged to enter the "not-polluting" business, and the government should not be paying companies for shutting down plants or going out of business altogether. Some companies might choose to install processes that generated excess waste because these might earn larger payments. The biggest subsidies would also tend to go to those firms that were doing the most polluting, while those that had done a better job of reducing pollution before the system went into effect would receive little or no reward. A subsidy program might also serve to increase total pollution by making it more attractive to go into the market covered by it.

Rather than paying companies to reduce pollution, a charge or tax could be levied on polluters. Theoretically this would encourage resources to be shifted toward less environmentally damaging products because it would add to the cost of goods that generated pollution or created waste. To be effective, charges should be on the amount of pollution generated rather than on end products. Otherwise there would be a lack of incentive to make or use the product in less damaging ways. There are a number of advantages to this approach. Those firms that found it the least costly to reduce pollution would be likely to make the largest reductions, thus efficiently allocating resources. Environmentally damaging products would be more costly, thus encouraging consumers to find substitutes. Because payments would be per unit of pollution, there would always be an incentive to reduce pollution levels even further, and polluters would not be restricted in what methods or technologies they might try. Charges would require those producing the pollution to pay for the damage, and the funds collected could be used for monitoring compliance and for supporting research to find better ways to reduce pollution.

Ideally, the level of any fee or tax should be exactly equal to the cost of the potential damage to society caused by the pollution or waste involved. Finding this proper level offers difficult problems, however. It is hard to estimate the value of aesthetics or of potential health problems. One approach would be to set charges in the hope of reaching a particular level of allowable pollution, and then the level could be increased until the target is reached. Fees should vary for different pollutants according to the potential harm they might cause to society. Charges also might vary by the time of day or year to help minimize the amount of pollution when atmospheric conditions are likely to make pollution worse. Varying fees by region would be dangerous, however, because it might encourage polluters to move their operations to where charges are lowest.

Emission permits are another approach to pollution abatement—allowing those facing the highest control costs to buy rights from others who can reduce emissions more cheaply. Once the system would be launched by a pollution control board, the rights might be bought and sold directly between companies. This would create a form of "property rights" for the use of air and water. Theoretically, as polluters sought to minimize their costs, resources would be efficiently allocated to achieve the maximum pollution reduction at the least cost. To start the process, permits might be allocated to existing polluters, or they might be auctioned to the highest bidders. Giving the rights away penalizes new firms, which are likely to make use of better control technologies. It would also deprive the government of a potential source of revenue. The giveaway approach would reward those who are polluting the environment. Thus an auction system seems preferable.

Many environmentalists oppose either a system of charges or permits because they see it as giving companies a "license to pollute." Some firms might choose to pay for their pollution rather than to install controls or waste-treatment facilities. Some critics feel that offering the option of paying for pollution would thus favor richer, larger companies while poorer ones would be forced to comply. In practice, however, if charges were to steadily increase, even richer firms would have to find ways to reduce their costs to remain competitive. Charges may be viewed as a form of fines, an approach commonly used by government to penalize speeders and tax evaders, for example. Some feel that incentives or penalties would not be as effective as regulations in reducing pollution. But, by using a market-based approach, there would always be an incentive to find more cost-effective methods for abatement and to reduce pollution even further than regulations would have required.

Policy shifts might also be accomplished by modifying or changing existing legislation and rules that actually encourage environmental deterioration. Tax benefits sometimes gained by filling in wetlands could be eliminated, for example, and violators of wetland regulations could lose their agricultural subsidies. If farmers in California could sell some of their water rights to cities, they might use the proceeds to install more effective water conservation systems for irrigation. Political clout has prevented many such changes in the past, but the growing momentum of the environmental movement will increasingly overcome the resistance of powerful pressure groups. In many cases, sometimes with the help of either existing or new technology, efficient ways can be found to support industrial and other development activities with much less damage to ecosystems. It is likely that many of these solutions will emerge from the efforts now being made to develop economic incentives or penalties encouraging better environmental stewardship.

THE OPPORTUNITIES

In summary, although the construction industry can expect to see steadily increasing pressures due to the environmental movement, the process is at the same time creating business opportunities in fast-growing, emerging markets. Even though the risks are higher and the dynamics quite different from the traditional construction business, there will be strong growth potential in numerous sectors, both domestically and abroad for decades to come. But those who enter these markets will have to be alert to changes in regulations and policies that may come out of shifts to market-oriented incentives and penalties and the quest for sustainable-development methods.

It will be important for every construction company to be alert to changes in the environmental sector because of the growing impact on an ever-widening span of its customer base. Clients will expect construction contractors to be informed, and those that become adept at giving worthwhile consultation and advice are likely to make strong competitive gains in the marketplace as a result.

APPENDIX A
MIT SYMPOSIUM ON THE GLOBAL ENVIRONMENT AND THE CONSTRUCTION INDUSTRY: ABSTRACTS OF PRESENTATIONS

SYMPOSIUM ABSTRACT

In recent years, concern for the quality of the environment has brought about fundamental and far-reaching changes in global construction markets. As firms, governments, and the public gain a better understanding of the scope and urgency of this concern, new ways to mitigate or avoid ecological damage will emerge and new markets will arise.

These markets will demand creative corporate strategies and more sophisticated management skills, novel technologies, and a new understanding of the environmental, social, legal, and economic contexts of industrial and construction activity. The recognition of these global markets and greater maneuverability across national borders provides new challenges and opportunities at the intersection of environment, industry, and construction.

The Symposium on Global Environment and the Construction Industry is intended to foster a better understanding of the implications of global environmental change on engineering and construction, and to identify emerging market opportunities. Hopefully we can come away from this symposium with appropriate and effective strategies for participating in this major market.

A diverse set of expertise from academia, industry, and government will participate in the symposium. Facility owners, engineering and consulting firms, contractors, and suppliers of technologies—all of whom share a concern for the environment—are invited to participate.

This symposium is being jointly sponsored by the Center for Construction Research and Education at MIT and Hazama Corporation of Japan.

Center for Construction Research and Education

The Center for Construction Research and Education (CCRE) was established in 1982 by the Department of Civil Engineering at MIT.

The Center's specific goals include the following:

- to develop, promote, and apply innovative technologiesand management methods that improve construction quality, productivity, and safety
- to create new mechanisms of university/industry partnership to stimulate technological and management innovations and to accelerate their transfer to industry
- to strengthen education programs by developing new graduate courses to anticipate the needs of construction industry leaders over the next nineteen to twenty years
- to develop a culture within the construction industry that recognizes and supports greater industry involvement in education and basic academic research
- to act as an agent among research, educational, and industrial efforts to improve the contribution of the U.S. engineering and construction industry to our economy and society

To achieve these goals, CCRE conducts education, research, and industry interaction programs that address fundamental issues of resources (capital, labor, and materials), technology (innovation, productivity, and automation), and management (at the industry, firm, and project levels).

Hazama Corporation

Hazama Corporation is one of Japan's leading construction companies. Founded over 100 years ago, the firm provides design and construction services on a wide variety of projects ranging from large-scale infrastructure projects such as roads, dams, bridges, tunnels, and power plants to building projects such as high-tech, high-quality skyscrapers and office buildings.

Hazama Corporation also has a long history of involvement in the international market including infrastructure construction in developing countries and, more recently, environmental projects in Eastern Europe. The company has a special concern for the social and economic impacts of environmental problems and the role that the international engineering and construction industry can play in preventing or mitigating these problems.

SESSION I:
GLOBAL ENVIRONMENT: CHALLENGES AND OPPORTUNITIES
Chair: **David H. Marks,** Head, Dept. of Civil Engineering, Massachusetts Institute of Technology

Policy Parameters for Environmental Protection: National and International Perspectives
NAZLI CHOUCRI
Professor, Political Science and Associate Director of the
Technology and Development Program, MIT

Nazli Choucri's areas of research revolve around international politics and global change, international energy problems, and population/energy/environment relationships. She is the author of *International Politics of Energy Interdependence* (1976) and *International Energy Futures* (1981); editor of *Multidisciplinary Perspectives on Population and Conflict* (1984); and editor of and contributor to *Global Change: Environmental Challenges and International Responses* (forthcoming).

Professor Choucri serves as a consultant to numerous agencies of the United Nations, firms with global markets, and various national and international scientific organizations. She heads the Middle East Program at MIT.

Global environmental issues are intensely political. The politicization of global change has already injected scientific evidence (and uncertainties) in the policy domain—national and international. It is the political process that will marshall social responses to global issues and ultimately legitimize the responses to evolving scientific evidence and concerns and corresponding technological options.

The dependence of policy making in this area on science, technology, and engineering is perhaps more pronounced than in other issues of national or international concern. Clearly the nation is a crucial institution in this regard: it remains the only legal jurisdiction enfranchised to act on behalf of citizens or to regulate their behavior. Regardless of the policy responses envisaged—and the role of industry, multinational corporations, and others—the institutions of the nation cannot be bypassed as significant actors.

Key input into policy formation regarding the natural record of global change must come from the sciences: key policy processes can only be understood in the context of analysis in the social sciences. The range of policy responses, or output, envisaged is influenced by prevailing applications of knowledge and skills and of engineering and technology.

All of these factors provide new parameters for policy and new challenges for policy analysis. This presentation addresses these new policy parameters.

The Changing Climate for Construction
WILLIAM R. MOOMAW
Director
Center for Environmental Management (CEM)
Tufts University

Dr. Moomaw is working primarily to develop policy options based upon available scientific knowledge that address human-induced global climate change, depletion of the stratospheric ozone layer, acid deposition, and air pollution. He is a physical chemist, having graduated from Williams College. He received his Ph.D. from MIT. Prior to coming to Tufts, he was Director of the Climate, Energy and Pollution Program at the World Resources Institute, Washington, D.C. For many years he was Fitch Professor of Chemistry and Environmental Studies and Director of the Center for Environmental Studies at Williams College where he won an outstanding teaching award. He has been a visiting scientist at UCLA, Tufts University, Wayne State University, and the Technical University of Munich. His research has been in the fields of laser spectroscopy, photochemistry, theoretical chemistry, and acid deposition. He was a Congressional Science Fellow of the stratospheric ozone depletion and U.S. energy and forestry policy during the 1970s. He was, for many years, chair of the American Chemical Society Task Force on Biotechnology and Toxic Substances and presently serves on the editorial board of several environmental journals. His paper on "Northwestern Responses to Global Climate Change" won a third prize in the International Mitchell Competition in 1991.

The threat of global climate change has the potential to alter building standards in a major way. European nations have already begun to tighten energy efficiency requirements from their already stringent standards in a first move towards lowering carbon dioxide emissions associated with space and water heating. Several states in the United States are also considering tighter building standards, and at least one has adopted them for public buildings under climate-change legislation. New technological innovations and materials permit more effective energy-conserving construction and alternative means of heating and cooling buildings than in the past. Lighting is another area in which major changes promise significant reductions in energy consumption and greenhouse-gas emissions.

On the supply side, the opportunities for incorporating solar technology into building design are also significant. In particular, the use of photovoltaic roofing shingles and other techniques for utilizing roofs as supports for photovoltaic electricity generation will be examined. Several proposals for mitigation and adaptation to climate change will also be discussed.

Environmental Markets: New Opportunities for Growth
JOHN R. EHRENFELD
Senior Research Associate and Director of MIT Program on Technology, Business, and the Environment
Center for Technology, Policy, and Industrial Development

Dr. Ehrenfeld has additional appointments at MIT as Lecturer in the Departments of Chemical Engineering and Urban Studies and Planning. At MIT since 1985, he coordinates the Hazardous Substances Management Program, an interdisciplinary educational, research, and policy program.

He is directing a major research project examining the way businesses manage environmental concerns, seeking organizational and technological changes to improve their practices.

In 1977 he was appointed by President Carter to serve as Chairman of the New England River Basins Commission (NERBC). There he was responsible for developing regional policies and strategies for surface and groundwater and coastal resources. He was instrumental in initiating and executing a major study of siting unwanted facilities, using power plants as the study topic.

Prior to his term at the NERBC, Dr. Ehrenfeld spent nineteen years performing and directing research, development, and consulting activities. Among many policy studies are a White Paper on significant deterioration due to air pollution that contributed to the Clean Air Act amendment process.

Dr. Ehrenfeld has served on the Massachusetts Water Resources Commission, the state's primary water policy organization. He is a member of the American Chemical Society, American Association for the Advancement of Science, Air Pollution Control Association, and Society for Risk Analysis; and is listed in American Men and Women of Science. He holds a B.S. and Sc.D. in Chemical Engineering from MIT.

The environmental industry has very old roots in the development of water supply and wastewater management and in air pollution control. Since the 1970s, the market for environmental goods and services has expanded rapidly to a size of about $80–100 billion today. This growth has been spawned largely by regulatory requirements in the United States and in other industrialized nations. In the coming decades, the global market will continue to grow rapidly as regulatory systems become even more established and also in response to private-sector initiatives answering the public's call for sustainability and environmentally sound processes and products.

SYMPOSIUM ABSTRACT

The history of environmental regulation will be traced, the current environmental industry will be described, and specific opportunities for and barriers to new environmental markets with emphasis on the construction industry will be discussed. While much recent focus has been placed on waste management and old chemical disposal site remediation, interesting opportunities may arise in transportation, water-resource development, and other projects related to the natural environment. Public concerns over virtually all construction projects require an enhanced political awareness in the industry.

SESSION II:
CORPORATE RESPONSE TO GLOBAL ENVIRONMENTAL CONCERNS
Chair: **Nazli Choucri,** Associate Director of the Technology and Development Program and Professor, Political Science Department, Massachusetts Institute of Technology

Implications of Industrial Ecology for Facility Design and Construction
R.A. LAUDISE
Director of Materials and Processing Laboratories
AT&T Bell Laboratories

Robert A. Laudise received his B.S. in Chemistry from Union College, Schenectady, New York, in 1952 and his Ph.D. in Inorganic Chemistry from MIT in 1956.

He is the Director of Materials and Processing Research Laboratory at AT&T Bell Laboratories, which is responsible for basic research in physical chemistry, the discovery of new electronic materials, the preparation of laser materials for optical communication, fundamental polymer studies, and materials characterization.

He joined AT&T Bell Laboratories in 1956. His interests at AT&T Bell Laboratories have included solid-state chemistry, materials science and materials conservation, and crystal growth. He has been especially interested in hydrothermal crystallization and in the preparation of piezoelectric, ferroelectric, nonlinear optical and related materials. He is Adjunct Professor of Materials Science at MIT.

Dr. Laudise is the author of the book, *The Growth of Single Crystals,* and more than 125 publications on crystal growth and related fields. He is the holder of twelve patents. He has received the A.D. Little Fellow (M.I.T), Sawyer Prize—1976, the International Crystal Growth Prize—1981, and the American Chemical Society Materials Chemistry Prize—1990. In 1989 the International Organization for Crystal Growth designated its prize for experimental crystal growth the Laudise Prize. He is a member of the International Organization of Crystal Growth (past president), 1978–83; the American Chemical Society (past Chairman of the Solid State Chemistry Sub-division); the National Academy of Engineering, 1980; the National Academy of Science, 1991; and a Fellow of the AAAS and the American Mineralogical Society.

Industrial ecology may be defined as a view of industrial products and processes that considers the total materials cycle from mine to junk in a way to eliminate or minimize adverse environmental effects. Industrial activities are designed to be a part of and synergistic with the larger natural ecosystem. The object is to get beyond remediation, beyond the belching smokestack, the dirty liquid-effluent pipe, beyond the contaminated landfill and to eliminate the root causes of environmental contamination from cradle-to-grave design and benign processing. The chemical and electronics industries are increasingly focusing their activities on products and processes whose industrial ecology is sound. The elimination of fluorochlorocarbons and the increasing cost of facilities for large-scale integrated circuit production is driving the electronics industry to new modes of circuit fabrication that will be described. The in-situ preparation of dangerous chemical feedstocks from benign precursors as they are used has promise for eliminating the need for the storage and shipment of dangerous reagents such as arsine. Recent progress in this area will be discussed. These trends will shape the sort of processes and hence the kinds of factories we will need in the twenty-first century. Too much factory construction, especially in the electronics industry, has been ad-hoc. Cost, yield, and ecological considerations are driving us towards a systems view of production facilities. Special emphasis will be placed on the current status of industrial ecology and the need for interaction with the design and construction sectors.

Opportunities and Challenges Related to Environment: A Government Perspective

DR. RAVI JAIN
Acting Director
Army Environmental Policy Institute

Ravi Jain founded the Army Environmental Policy Institute. An environmental engineer and R&D manager for 25 years, Dr. Jain received degrees in Civil Engineering (B.S., MSCE) from California State University. He earned a Ph.D. at Texas Tech University and a Masters in Public Administration from Harvard University. A research affiliate of the Massachusetts Institute of Technology, Dr. Jain is also an Adjunct Professor of the University of Illinois at Urbana-Champaign. He has served as Chairman of the Environmental Engineering Research Council, ASCE, and is a member of the American Academy of Environmental Engineers and a Fellow of ASCE.

His many awards and honors include the Army's highest research award: the Army R&D Achievement Award and the Army Decoration for Meritorious Civilian Service. He was also a recipient of the Founder's Gold Medal and named "Federal Engineer of the Year" for 1989. He has coauthored or edited seven books.

Introducing an external force, such as environmental regulations, in a free market economy is likely to disrupt its efficiency. It is generally believed that perfectly competitive markets, devoid of such regulations, can produce efficient (Pareto optimal) outcomes. Environmental regulations, thus, can affect the cost and completion time of construction and other projects.

The challenges that we face are to first understand public-policy issues related to these regulations. These public-policy issues deal with items such as: basis for promulgating environmental regulations, economics of regulations, and the public view of environmental regulations. Understanding these issues can provide a clearer perspective as to why industrialized countries are taking steps to establish environmental regulations. Related to these public-policy issues are also many concerns regarding these regulations. These concerns deal with the economics of regulations and their effect on industrial productivity.

There are indeed a number of opportunities that result from these challenges. There are certain inefficiencies in the ways that these regulations are promulgated and implemented. One challenge is to develop strategies to overcome these inefficiencies. In addition, environmental issues present other opportunities for undertaking innovative construction projects. For example, construction projects are required to build treatment facilities and processes to comply with environmental regulations. Billions of dollars are expected to be spent in the United States for Super-fund site cleanup. Global opportunities will also emerge for the construction industry in dealing with the crucial environmental problems facing Eastern Europe, the former Soviet Union, and other emerging industrial nations. Solutions to global problems such as climate change will present other significant opportunities and challenges for the construction industry.

Global Environment and the Electrical Power Industry: A Period of Conflict, Growth, New Supply, and End-Use Technology
DR. DAVID C. WHITE
Ford Professor of Engineering
Department of Electrical Engineering and Computer Science
Massachusetts Institute of Technology

As Founding Director of the Energy Laboratory at MIT in 1972, where he served as Director until 1988, David C. White was involved in its growth into a major interdisciplinary research laboratory.

Professor White attended Stanford University, where he received the B.S., M.S., and Ph.D. in Electrical Engineering in the years 1946, 1947, and 1949, respectively. He was a fellowship student at Stanford University from 1946 through 1949 and then joined the faculty of the University of Florida as an Associate Professor from 1949 to 1952. In 1952 he was appointed to the Faculty of Electrical Engineering at MIT as an Assistant Professor; promoted to Associate Professor in 1954; Professor of Electrical Engineering in 1958; Ford Professor of Engineering in 1962; and Director of the Energy Laboratory, 1972–88. His publications include a number of articles in professional journals and the textbook *Electromechanical Energy Conversion* (Wiley, 1959).

Dr. White is a member of the National Academy of Engineering, the Institute of Electrical and Electronics Engineers (Fellow), the American Society for Engineering Education, the American Academy of Arts and Sciences (Fellow), and the Honorary Societies Phi Beta Kappa, Sigma Xi, Tau Beta Pi, and Eta Kappa Nu.

Professor White is a consultant and energy advisor to several industrial firms.

Technology change in both the supply and end use of electrical power has been a natural part of the evolutionary growth of electrical systems in meeting the needs of industrialized societies. This change was driven first by technical innovation toward more efficient prime movers, electrical-supply equipment, and electrical-consuming equipment. By the middle of the twentieth century, regional environmental concerns about sulfur oxides (SO_x), nitrogen oxides (NO_x), and particulate emissions to the atmosphere, thermal emission to air and water, and liquid and solid pollutants to the water and soil brought forth new technology improvements. Global environmental concerns, particularly global warming, are growing in importance at the end of the twentieth century and will probably last well into the foreseeable future. These concerns increase the incentive to improve further all electrical-supply and end-use equipment and systems.

Today there is no clear consensus among concerned social groups about the future direction of electrical energy systems or, for that matter, any of the various energy-supply and end-use technologies. The best way to characterize the present dialogue on electric power systems is that specific groups (whether they be environmentalists, technologists, conservationists, utilities, and so on) tend to focus on one-dimensional solutions to multidimensional problems. As a result, simplistic solutions are proposed that do not lead to constructive social action.

In this presentation, we take the position that history shows that technological innovation has worked well in the past, and there is ample opportunity for it to deal effectively with societal concerns in the future. However, this requires taking a long-term view that focuses on the total energy system. As such, electricity use could be either a larger or a smaller percentage of overall energy use. Considering all social, environmental, and economic factors, we believe there is solid evidence that electricity will be a growing choice for energy supply in the future. This will require major changes in the way electrical energy systems are planned, financed, and operated.

Since such change must be evolutionary, it cannot be accomplished by a series of short-term, narrowly focused technical fixes. It is impossible for these changes to occur only in one part of the globe (e.g., the United States) Global warming has made us sensitive to the planet earth as a truly coupled system. World events of the past year, specifically the events of August 1991, point to an even closer coupling of world economics and social systems.

We will discuss possible directions which the electrical energy system may take and the potential opportunities for worldwide construction of electrical energy systems. The opportunities are very large; unfortunately so are the constraints. In the wise words of Pogo: *"We have found the enemy and he is us."*

SESSION III:
DEMAND FOR TECHNOLOGICAL INNOVATION
Chair: **Howard B. Stussman,** Editor-in-Chief, *Engineering News Record*

Innovative Technologies, Where Are They?
HOWARD B. STUSSMAN
Editor-in-Chief
Engineering News Record

Prior to joining *Engineering News Record* in 1965, Mr. Stussman was a police reporter, court reporter, and the Assistant City Editor for *The News*, a daily newspaper in Los Angeles. He also worked as a copy editor for the *Los Angeles Times*.

He came to *Engineering News Record* as an assistant editor in the magazine's construction economics department. A year later, he joined the United States Naval Mobile Construction Battalion Ten as a journalist, shipping out twice to Vietnam. During that time he wrote numerous articles for the Navy and bylined articles for *Stars and Stripes* and for newspapers in the United States, mainly on the construction effort in Vietnam. He received a number of citations and commendations for his writing during this period.

On returning to *Engineering News Record,* Mr. Stussman was appointed Assistant Managing Editor and has held subsequent writing and management positions in *Engineering News Record's* management and labor, water and power, transportation and buildings departments. he has traveled extensively for *Engineering News Record*—including to Europe, the Far East, and the Middle East—writing about companies, dams, tunnels, buildings, and highway projects.

Prior to becoming Editor-in-Chief, he was the magazine's Executive Editor and Managing Editor, responsible for editorial and financial planning, coordination of special editorial projects, and personnel administration. He planned and implemented the computerization of the magazine's editorial operations, from statistical processing to worldwide communications.

Mr. Stussman holds a Bachelor of Science degree in Journalism from California State University, Northridge, and has completed numerous business-writing and executive-management courses.

The road to new and innovative environmental cleanup technologies is littered with good intentions. In creating the Resource Conservation and Recovery Act, the 1984 Hazardous and Solid Wastes Amendments, the Comprehensive Environmental Response, Compensation, and Liability Act, and the 1986 Superfund Amendments and Reauthorization Act, the collective legislative heart was in the right place, but the mind obviously was involved in some sort of out-of-body experience.

That, at least, would explain the existence of this Sargasso Sea of weedy environmental laws, gaseous regulations, and guidance documents and why so many engineers and constructors find commerce in this area hazardous to one's corporate health.

It also explains the lack of development of much-needed new technologies in the face of the founding of a Federal Technology Innovation Office, staffed by some of the nation's best and brightest in the environmental services field.

Getting out of the innovation doldrums requires legislation to prime the technology pumps with incentives and liability protection and actions that will clear away the technical, regulatory, and institutional barriers blocking innovative technologies from market.

Innovative Site Remediation Technologies: Opportunities and Barriers
WALTER KOVALICK
Director
Technology Innovation Office
Environmental Protection Agency

Dr. Kovalick manages an office chartered to act as a champion for the introduction of more innovative remediation technologies in the cleanup of abandoned waste sites and Superfund and corrective actions under the Resource Conservation and Recovery Act. Formed in 1989, his office is providing policy leadership and technology information within the EPA and enabling the broader use of innovative technologies through other federal agencies, states, consulting engineers, technology vendors, and other countries.

For the previous five years, Dr. Kovalick was the Deputy Director of the Superfund program where he shared leadership responsibilities for a nationwide program to respond to hazardous substance releases—both of an emergency nature as well as from abandoned waste disposal sites.

Joining the EPA in 1970 from one of its predecessor agencies, Dr. Kovalick worked in EPA's headquarters and two of its regional offices to develop an effective partnership with the states in implementing the Clean Air Act. From 1974 to 1978, he managed a program to develop the initial standards to define hazardous wastes and to regulate generators and transporters of such waste. Until 1984, he directed a staff office in the Office of Toxic Substances to develop strategies and information systems for agencywide regulation of chemicals. He has represented the EPA on hazardous waste and chemical issues to several international organizations, served as a consultant to the United Nations Environment

Program and worked for the past five years on a NATO project to share information on remediation technologies.

Dr. Kovalick has a Bachelor of Science in Industrial Engineering and Management Science from Northwestern University and a Masters of Business Administration from Harvard Business School. He has a Ph.D. in Public Administration from Virginia Tech. He is a recipient of the EPA Bronze and Silver Medals for Superior Service. He is a member of the American Society for Public Administration, the Institute for Industrial Engineers, the Association of Public Policy and Management, and the Academy of Management.

The United States and several other industrialized nations have made public-policy decisions to begin to remediate contaminated soil, abandoned waste sites, and the groundwater beneath such sites. While these programs were begun about a decade ago, traditional technology solutions have dominated the cleanup marketplace. Innovative technology development is in the very nascent stages of the classical technology development process—proof of concept, bench and pilot-scale testing, and full-scale commercialization.

The barriers to such technology development and application in the United States include information transfer and use, regulatory hurdles, and institutional impediments. Contraposed to these problems is the assessment of a large market for less expensive, more cost-effective solutions as well as governmental initiatives to enable further technology commercialization. While engineering expertise has been manifest in the early stages of site characterization and feasibility studies, the U.S. construction engineering community has yet to reach its full potential for its engineering contribution to this important issue.

Regulation-Driven Versus Market-Dictated Technologies
FRED MOAVENZADEH
Director
Center for Construction Research and Education and
George Macomber Professor of Construction Management
Massachusetts Institute of Technology

Dr. Moavenzadeh's professional field of interest is construction engineering and management, with a primary focus on international construction, construction finance, and strategic management. He has taught the basic courses in construction, facility design, and engineering and management of infrastructures, both in the Department of Civil Engineering at MIT and at the Graduate School of Design at Harvard University.

Over the past twenty-five years, Dr. Moavenzadeh has directed a series of research programs relating to construction engineering and management, both in the United States and in developing countries. These include international construction finance; merger and acquisition in the construction industry; nature and organization of construction in the United States; and comparative studies of U.S. construction with that of Europe, Japan, and Korea. Most recently he conducted a major study on the globalization of construction firms and the need for restructuring the construction industry in light of recent changes in the global market and new developments in the information and communication fields.

Dr. Moavenzadeh has served as a consultant to several U.S. and international agencies, including the International Bank for Reconstruction, the Inter-American Development Bank, the United Nations Industrial Development Organization, the United Nations Center for Human Settlements, and the U.S. Agency for International Development. He has served on U.S. and international panels relating to construction engineering and management and has authored over 100 professional papers.

SYMPOSIUM ABSTRACT

Conventional government approaches to pollution control rely almost exclusively on command-and-control regulations that either specify the technology that must be used (technology-based standards) or set an emission rate cap that all sources must meet (uniform performance standards). In either case, the innovation of new pollution control technologies is stifled by a market that is driven by the regulations that require their use rather than the needs of the consumer that will pay for them.

If future policy approaches are to be effective in achieving environmental objectives, they must first ensure that consumers and producers face the true (both direct and social) costs of their inefficient natural-resource use and environmental degradation. Then, new policies must include market-based mechanisms that will create economic incentives for exceeding pollution-control standards. In this way, industry will be encouraged to develop new technologies that both save the environment and offer new potential for increased profits.

Five general categories of policy instruments that are moving in this direction include: pollution charges; tradable-permit systems; deposit-refund systems; removal of government-mandated barriers to market activity; and elimination of government subsidies of environmentally unsound practices.

No single one of these mechanisms can achieve the nation's environmental goals alone. Command-and-control regulations will still be necessary as will the EPA's continuating enforcement role. However, the balanced use of a variety of policy options that enlist the forces of the marketplace and the ingenuity of the entrepreneur will result in more rapid development of new technologies and, ultimately, a more rapid solution to our environmental problems.

SESSION IV:
INNOVATIVE TECHNOLOGIES
Chair: **Fumio Sugimoto,** Manager, Corporate Planning Division, Hazama Corporation, Tokyo, Japan

Trends in Design of Fossil-Fueled Power Plant: Response to Environmental Constraints
ADEL F. SAROFIM
Lammot Dupont Professor of Chemical Engineering
Massachusetts Institute of Technology

Dr. Sarofim has held his present position at MIT since 1989. He received his Sc.D. in Chemical Engineering at MIT in 1962 and has been affiliated with MIT since 1960 as an Instructor in Chemical Engineering; Assistant Professor in Chemical Engineering, 1961–67; Associate Professor, 1967–72; and full Professor in Chemical Engineering since 1972.

Dr. Sarofim has been a Visiting Professor at Sheffield University, England; the University of Naples, Italy; and at the California Institute of Technology. He has served on numerous national committees including the New England Consortium on Air Pollution Board of Directors, 1972–74; National Academy of Sciences Committee on Chemicals in the Environment, 1972–74; American Flame Research Committee Technical Secretary, 1967–76; H.E.W. Committee on Health and Ecological Effects of

Increased Coal Utilization, 1977; Health Effects Working Group on Coal Technologies, Interagency Committee on the Health and Environmental Effects of Energy Technologies, 1979; the American Association of Aerosol Research, Board of Directors, 1981–86; National Research Council, Energy Engineering Board, 1983 to the present; National Research Council, Committee on Chemical Engineering Frontiers, 1984–88; and the EPA Science Advisory Board, Strategic Research Sub-Committee, 1987–88.

In addition, Professor Sarofim has served on several editorial advisory boards, including the *Solar Energy Journal, International Journal of Heat and Mass Transfer, Progress in Energy and Combustion Science, Combustion Science and Technology,* and *Aerosol Science and Technology.*

He has been awarded the Sir Alfred Edgerton Gold Medal from the Combustion Institute in 1984; Kuwait Prize for Petrochemical Engineering, 1983; Hoyt C. Hottel Lecturer, Combustion Institute, 1986; and Lacey Lecturer, California Institute of Technology, 1987. Dr. Sarofim is the author and coauthor of over 150 papers.

The major changes that have occurred during the past decade in the design of fossil-fueled power plants have been driven by the need to reduce the emission of the acid-rain precursors, NO_x and SO_x. The technologies adopted have included a mix of combustion-process modification and the installation of flue-gas treatment systems, with major inroads being made recently by fluidized-bed combustors. The emerging concerns with global climate change have added nitrous oxide (N_2O) and carbon dioxide (CO_2) to the gases that need to be controlled, and this has led to the development of systems involving pressurized combustion and the use of a combination of a gas turbine with a steam turbine. The presentation will briefly review the changes in technologies, the status of their development, and their impact on emissions.

Innovative Technologies—The Challenges of Implementation
PETER D. ARROWSMITH
President and Chief Operating Officer
HALLIBURTON NUS Environmental Corp.

Peter D. Arrowsmith's company was formed in July 1991 through the consolidation of two Halliburton companies—NUS Corporation and Halliburton Environmental Technologies, Inc. Prior to assuming similar responsibilities at NUS, Mr. Arrowsmith managed the energy service activities and architect engineering—nuclear design operation. Before joining NUS as a projects manager in 1963, he was assigned to the Atomic Energy Commission's Naval Reactors Division where he was involved in the construction and testing of nuclear-powered submarines.

Mr. Arrowsmith holds a B.S. in Chemical Engineering from the University of Kansas and advanced training at the Oak Ridge School of Reactor Technology and the Wharton Business School.

Although it is becoming inceasingly clear that new and innovative remediation technologies will be required to efficiently and cost-effectively clean up thousands of toxic waste sites in the United States, institutional barriers often severely impact the timely implementation of such innovation.

We will discuss the substantial institutional barriers together with the current regulatory, contracting, and risk-sharing regimes that contribute to delay in such implementation, as well as look at specific regulatory/contracting changes and financial incentives that could contribute to the acceleration of innovation.

SYMPOSIUM ABSTRACT

Geothermal Power Systems as an Example of Alternative Energy Technologies: Their Development Status and Impact on the Construction Industry

JEFFERSON W. TESTER
Director of the MIT Energy Lab
Professor of Chemical Engineering
Massachusetts Institute of Technology

Dr. Tester has been involved in various aspects of chemical engineering process research as it relates to energy technology and environmental-control technology for the past nineteen years. For the last eleven years, Dr. Tester has been active in examining the oxidation of toxic compounds in supercritical water, both at MIT in his research program and as a consultant to various laboratories and companies.

He has coauthored more than seventy papers on various topics in applied thermodynamics, kinetics, and transport; and four books related to energy and environmental issues ranging from geothermal reservoir and drilling technology and power-conversion system design and economics, to assessing local, regional, and global environmental effects caused by energy supply and use. He currently teaches graduate and undergraduate subjects in thermodynamics and has won several teaching awards. He was selected as the 1980 Outstanding Chemical Engineer by the New Mexico Chapter of the American Institute of Chemists.

Dr. Tester received his B.S. and M.S. from Cornell University in Chemical Engineering and his Ph.D. from MIT. He is a member of the American Institute of Chemical Engineers, American Chemical Society, the Society of Petroleum Engineers, and the Geothermal Resources Council. He has served as an advisor to the DOE and the National Research Council in areas related to geothermal energy technology and waste minimization and pollution reduction.

Geothermal energy, which utilizes the natural heat contained in the earth's crust, can provide a widely available source of nonpolluting energy. For example, it could help mitigate the continued buildup of carbon dioxide in the atmosphere—one of the major environmental consequences of our ever-increasing use of fossil fuels for heating and power generation. The earth's heat represents a well-distributed and almost unlimited source of energy that can begin to be exploited within the next decade through the hot dry rock (HDR) heat-mining concept being actively developed in the United States and in several other countries. Geothermal energy can be used directly to produce power or indirectly in hybrid geothermal/fossil-fueled systems in diverse applications such as:

- baseload electric power generation
- direct heat use
- cogeneration
- feedwater heating in conventional power plants
- pumped storage/load-leveling power generation

The nature of geothermal resources and the technology required to implement the heat-mining concept in several applications will be discussed. An assessment of the drilling and reservoir stimulation requirements for establishing hydrothermal and HDR feasibility will be presented in the context of providing a commercially competitive energy source.

Subsurface reservoir and surface power-plant designs and performance issues will be discussed in the context of estimating the materials, downhole and energy conversion equipment, and financial resources required for geothermal energy to have a significant impact on worldwide energy supply in the twenty-first century.

SESSION V:
STRATEGIC RESPONSE TO GLOBAL ENVIRONMENTAL ISSUES
Chair: **Fred Moavenzadeh,** Director of CCRE and George Macomber Professor of Construction Management, Massachusetts Institute of Technology

The Effects of Global Environmental Concerns on the Engineering Industry
ROBERT C. MARINI
President and Chief Executive Officer
Camp Dresser & McKee, Inc. (CDM)

Robert C. Marini's company is one of the world's leading environmental engineering firms. He has played an active role in CDM's domestic and international practice since joining the firm in 1955. Milestones at CDM include becoming a partner in 1967, Executive Vice President in 1982, and President in 1984, with CEO and Chairman added to his responsibilities in 1989. A registered professional engineering in twenty-one states, Marini holds a Bachelor's Degree in Civil Engineering from Northeastern University and a Master's Degree in Sanitary Engineering from Harvard University.

Active in many professional endeavors, Mr. Marini was elected in February 1991 to the National Academy of Engineering, one of the highest professional distinctions accorded to an engineer. He also is a Diplomate of the American Academy of Environmental Engineers; a Fellow of the American Society of Civil Engineers and the American Consulting Engineers Council; and a member of numerous engineering associations, including the American Water Works Association, Water Pollution Control Federation, and the International Association of Water Pollution Research and Control. Mr. Marini's professional accomplishments range from being named "Young Engineer of the Year" in 1966 by the Massachusetts Society of Professional Engineers, to receiving the Distinguished Eagle Scout Award in 1986 from the National Council of the Boy Scouts of America.

The engineering industry has long worked as a partner with the construction industry in designing and building large-scale facilities for public and private clients. A major current focus of environmental engineering is how to clean up solid and industrial waste that was improperly disposed of in the past or that are being generated today. Another focus is evaluating and mitigating the impacts of new siting and building activities. Heightened concern for the global environment will greatly change the way we do business in two ways: there will be more attention to generation at the source, in addition to waste management; and there will be a different mix of facility types being constructed, with an increased focus on environmental protection and energy and resource conservation.

First, greater attention to source control will mean more engineering involvement in the manufacturing process. Our clients will become partners in our work as we turn our attention to in-plant improvements, choices of materials, types of products to produce, and end-of-pipe results. This raises new questions about how we organize our work and who we hire to do it. Will we become more involved in proprietary technologies, which we've often shunned in the past? And will we need to do more turnkey operations, where engineer and contractor become partners, thus blurring the traditional lines of communication and responsibility? Environmental engineering expertise will have to expand from the traditional civil engineering arena to include greater expertise in process engineering, which means bringing in new talents and new people.

Second, the shift in the mix of constructed facilities to address environmental protection and resource conservation will require better design and tighter management controls. Both will be very dependent on enhanced computer-based data management and visualization of the design/construction process. Learning the necessary skills, fostering professional growth in these areas, and expanding our computer expertise and resources will not only tax our abilities—it will also affect how we price our services and where we make strategic alliances.

Environmental concerns are no longer the realm of the grassroots activist. As these issues have grown to become everyone's issues—truly global issues—the engineering and construction industries will need to evolve and work closer together to foster environmental protection and energy and resource conservation in the design and construction of facilities worldwide.

Developments in Europe: Opportunities and Constraints
BRADFORD S. GENTRY, ESQUIRE
Morrison & Foerster

Bradford Gentry, a partner in Morrison & Foerster and resident in the firm's London office, advises multinational manufacturing companies on environmental issues facing their operations in different countries. He also works with companies offering pollution control goods and services.

Mr. Gentry received his law degree, magna cum laude, from Harvard Law School. He is a Lecturer in Comparative Environmental Law at King's College, University of London and has frequently spoken and written on international environmental issues.

In the past three years, environmental issues have emerged as major public and political concerns throughout Europe—west, central, and east. The legislative response has been striking, with the adoption of many much more stringent environmental requirements. In addition, European consumers are increasingly demanding that products be designed and manufactured in an environmentally sensitive manner.

These trends pose clear opportunities for the construction industry. For example, a recent report estimated that over $1 trillion will be spent during the 1990s to address environmental issues in Western Europe alone. Opportunities are presented in a wide range of areas, including wastewater treatment; air pollution control; higher-efficiency energy plants; contaminated land cleanup, and hazardous waste disposal.

At the same time, there are a number of hurdles that face efforts to capture some of these opportunities. Some affect particular sectors: for example, waste disposal costs in southern Europe are so low and public opposition to new disposal facilities so high that surprisingly it is difficult to develop the new state-of-the-art incineration/treatment facilities that are critically needed. Others affect particular regions: for example, throughout central and eastern Europe, the need for environmental projects is clear, but the sources of convertible funds for such projects can be obscure. Still others extend to Europe as a whole, particularly given the range and diversities of cultures and needs involved.

All in all, while there will be relatively few quick hits, major opportunities will continue to exist in the European environmental sector for the discerning company.

Global Environmental Issues and the Response of the Japanese Construction Industry
AKIRA KAGAMI
Vice President
Hazama Corporation and Chairman of the
Environmental Committee of the Japan Civil
Engineering Contractors' Association, Inc.

Mr. Akira Kagami graduated from the Department of Civil Engineering at Yamanashi Technical Institute, now Yamanashi University, in 1949. He then joined Hazama Corporation and has held the positions of Project Manager, Deputy General Manager, Director and General Manager of the Sendai Branch, Managing Director and Deputy General Manager in the Business Development Division, and General Manager of Hazama's Civil Engineering Business Development Division. Since 1990 he has served as Vice President and Representative Director of Hazama Corporation.

In addition, Mr. Kagami has played key roles throughout his career in a number of outside organizations including the Japan Civil Engineering Contractors' Association, Japan Electric Power Constructors' Association, Japan Railway Contractors' Association, Japan Road Contractors' Association, Japan Federation of Construction Contractors, Japan Ocean Development Construction Association, and Japan Society of Civil Engineers. One of the important roles he is playing currently includes that of Chairman of the Environmental Committee of the Japan Civil Engineering Contractors' Association, Inc.

The Japanese construction industry played a key role in the massive economic reconstruction efforts that took place following World War II. These efforts, however, as well as the two oil crises of the 1970s caused numerous environmental problems for Japan, and the government embarked on an environmental preservation program to save its energy and natural resources.

Despite the fact that environmental preservation became an ethic in Japan in the 1980s, the environment was still in danger from construction waste. In addition, Japan found itself exposed to global environmental problems, including warming gas.

Japan attacked these problems multilaterally. Among others, the construction industry is responsible for the future of the earth. The construction industry has the ability to balance nature and technology, and the Japanese construction industry and the Japan Civil Engineering Contractors' Association have been working together to try and solve global environmental problems.

Success in tackling global environmental problems is based on an "internationally cooperative approach," and Japan is working towards this goal.

APPENDIX B
SURVEY QUESTIONNAIRE: CONTRACTORS, DESIGNERS, AND TECHNOLOGY VENDORS

SURVEY QUESTIONNAIRE: CONTRACTORS, DESIGNERS, AND TECHNOLOGY VENDORS

SURVEY QUESTIONNAIRE
Section I: General Information about Your Firm

The following questions are intended to provide general information about your firm and the types of work that you perform. This general information will be used to categorize your company within the total survey sample, thus allowing comparisons and contrasts with other firms. The specific questions asking for your name and company name are optional. This information would be very helpful for tracking your response to this questionnaire, but it will not be used to identify you, your company, or your responses to any of the questions.

1. When was your company formed?

    a. before 1950 ___   b. 1950 - 1969 ___   c. 1970 - 1979 ___
    d. 1980 - 1989 ___   e. after 1989 ___

2. What is the size of your company?

    a. annual gross revenue* _____

    *Note: Please indicate if you report a figure besides annual gross revenue. For example, many design firms report total annual billings as a measure of financial size.

    b. number of employees: 1-10 ___   11-100 ___   101-1000 ___   >1000 ___

3. For the purposes of the survey, it is necessary to determine your company's role in the construction industry. You may be a client or supplier of the industry rather than a constructor or designer. Please use the following definitions to classify your company.

    owner: owns constructed facilities (e.g., oil companies, utilities)
    constructor: undertakes construction projects (i.e., a builder)
    designer: designs construction projects (e.g., architects, engineers)
    technology vendor: develops and/or sells technology for environmental field
    supplier: supplies materials, equipment, etc., for construction

    With regards to the construction industry, please indicate the principal role of your company in its regular business activities. If your company fills several of these roles, please mark all that apply.

    a. public owner ___
    b. private owner ___
    c. engineering or design firm ___
    d. technology vendor ___
    e. supplier ___
    f. constructor: general contractor ___   specialty (sub) contractor ___

SURVEY QUESTIONNAIRE: CONTRACTORS, DESIGNERS, AND TECHNOLOGY VENDORS

4. Is your company a subsidiary of another company?

                Yes ____   No ____

5. Please indicate the extent of your company's involvement - measured as a percentage of total business activity - in the following areas.

   *Note: Not at all = 0%   Minor = < 10%   Modest = 11-30%
               Major = 30-99 % Exclusive = 100%

	Not at all	Minor	Modest	Major	Exclusive
a. Petrochemicals	1	2	3	4	5
b. Water Supply	1	2	3	4	5
c. Sewer / Waste Water	1	2	3	4	5
d. Solid Waste	1	2	3	4	5
e. Hazardous Waste	1	2	3	4	5
f. Energy	1	2	3	4	5
g. Pollution Abatement	1	2	3	4	5
h. Environmental Assessment	1	2	3	4	5

6. Your company headquarters is located in what country?

   _____

7. Please indicate the geographical scope of your company by giving the percentage of your work which is:

   a. local ____% b. regional ____% c. national ____% d. international ____%

8. Please indicate the extent of your company's operations in each of the listed global regions. (Measured as a percentage of total business operations.)

   Note: Not at all = 0%   Minor = < 10%   Modest = 11-30%
              Major = 30-99 % Exclusive = 100%

	Not at all	Minor	Modest	Major	Exclusive
a. North America	1	2	3	4	5
b. West Europe	1	2	3	4	5
c. East Europe	1	2	3	4	5
d. Latin America	1	2	3	4	5
e. Far-East	1	2	3	4	5
f. Mid-East	1	2	3	4	5
g. Other _____	1	2	3	4	5

SURVEY QUESTIONNAIRE: CONTRACTORS, DESIGNERS, AND TECHNOLOGY VENDORS

Questions 9 & 10 are for constructors and designers only.

9. What percentage of your work is public? What percentage is private?

   Public _____ %   Private _____ %

10. Please indicate the frequency of the following contracting methods for the work (both public and private) undertaken by your firm?

	Never	Not Often	Often	Very Often	Always
a. competitive bidding	1	2	3	4	5
b. cost plus fee	1	2	3	4	5
c. negotiated price	1	2	3	4	5
d. other [please specify: _____]	1	2	3	4	5

SURVEY QUESTIONNAIRE: CONTRACTORS, DESIGNERS, AND TECHNOLOGY VENDORS

Section II: The Environmental Market

Many dimensions of the rapidly emerging environmental market offer promise for the engineering and construction industry. For the purposes of this survey, the environmental market will include business opportunities derived from the following areas of global environmental concern:

Acid Rain / Ozone Depletion / Global Warming, Clean Air, Deforestation, Energy, Solid Waste Management, Water Supply, Sewage / Waste Water Treatment, Environmental Assessment Technologies, and Hazardous Waste Remediation and Management.

This section is designed to assess your general views of this market, including its opportunities and challenges.

A. Market Size, Location, and Type

1. In your estimate, what is the approximate <u>annual</u> size (in current US dollars) of the total environmental market (defined above in the introduction to Section II):

    i. in your country (meaning the country of your company's headquarters) today?
       a. < 1 billion ___   b. 1-10 billion ___   c. 11-50 billion ___
       d. 51-100 billion ___ e. > 100 billion ___

    ii. worldwide today?
       a. < 1 billion ___   b. 1-10 billion ___   c. 11-50 billion ___
       d. 51-100 billion ___ e. > 100 billion ___

    iii. in your country (meaning the country of your company's headquarters) in the year 2000?
       a. < 1 billion ___   b. 1-10 billion ___   c. 11-50 billion ___
       d. 51-100 billion ___ e. > 100 billion ___

    iv. worldwide in the year 2000?
       a. < 1 billion ___   b. 1-10 billion ___   c. 11-50 billion ___
       d. 51-100 billion ___ e. > 100 billion ___

2. Between now and the year 2000 do you expect the environmental market to:
    a. shrink ___   b. stabilize ___   c. grow ___

    If you expect it to <u>grow</u>, at what rate:

Slowly		Moderately		Very Quickly
1	2	3	4	5

SURVEY QUESTIONNAIRE: CONTRACTORS, DESIGNERS, AND TECHNOLOGY VENDORS

3. Please indicate the relative (compared to each other) <u>current</u> environmental market sizes in these regions of the world.

	Small		Medium		Large
a. North America	1	2	3	4	5
b. West Europe	1	2	3	4	5
c. East Europe	1	2	3	4	5
d. Latin America	1	2	3	4	5
e. Far-East	1	2	3	4	5
f. Mid-East	1	2	3	4	5
g. Other	1	2	3	4	5

4. Please indicate the potential opportunity over the next five years - in the country of your company's headquarters - for the construction industry in the following areas:

	Minor		Modest		Major	Don't Know
a. Petrochemicals	1	2	3	4	5	9
b. Water Supply	1	2	3	4	5	9
c. Sewer / Waste Water	1	2	3	4	5	9
d. Solid Waste	1	2	3	4	5	9
e. Hazardous Waste	1	2	3	4	5	9
f. Energy	1	2	3	4	5	9
g. Pollution Abatement	1	2	3	4	5	9
h. Environmental Assessment	1	2	3	4	5	9

B. Regulatory and Legal Issues

5. How would you rate the current environmental regulation in the following areas?

	Excessive	Not Excessive	Don't Know
a. Petrochemicals	___	___	___
b. Water Supply	___	___	___
c. Sewer / Waste Water	___	___	___
d. Solid Waste	___	___	___
e. Hazardous Waste	___	___	___
f. Energy	___	___	___
g. Pollution Abatement	___	___	___
h. Environmental Assessment	___	___	___

SURVEY QUESTIONNAIRE: CONTRACTORS, DESIGNERS, AND TECHNOLOGY VENDORS

6. To what degree should the environmental market be driven by regulation versus market forces?

100% Reg.	75% Reg.	50% Reg.	25% Reg.	0% Reg.
0% Mar.	25% Mar	50% Mar.	75% Mar.	100% Mar.
1	2	3	4	5	6	7	8	9

7. To what degree should new regulations be punitive versus incentive based?

100% Pun.	75% Pun.	50% Pun.	25% Pun.	0% Pun.
0% Incen.	25% Incen.	50% Incen.	75% Incen.	100% Incen.
1	2	3	4	5	6	7	8	9

8. How important will the following levels of government be in determining the environmental regulations of the future?

	Not Important		Moderately Important		Very Important
a. local	1	2	3	4	5
b. state	1	2	3	4	5
c. federal	1	2	3	4	5

9. a. With regard to <u>environmental</u> regulations, how important is it to establish international regulations?

Not Important		Moderately Important		Very Important
1	2	3	4	5

   b. In the absence of international regulations, how important is it for neighboring nations to have comparable environmental regulations?

Not Important		Moderately Important		Very Important
1	2	3	4	5

10. a. What is the likelihood that future environmental regulations will result in additional work for the construction industry?

Low Probability		Moderate Probability		High Probability
1	2	3	4	5

    b. What is the likelihood that future environmental regulations will result in greater restrictions on construction?

Low Probability		Moderate Probability		High Probability
1	2	3	4	5

SURVEY QUESTIONNAIRE: CONTRACTORS, DESIGNERS, AND TECHNOLOGY VENDORS

11. In your opinion, is potential liability from environmental work a significant hindrance to operating in the environmental market?

Not Significant		Significant		Very Significant
1	2	3	4	5

12. When compared to other construction markets, to what degree are the following institutions wary of the environmental market because of potential liability issues?

	Less Wary		Same		More Wary
a. lending companies	1	2	3	4	5
b. insurance companies	1	2	3	4	5
c. bonding companies	1	2	3	4	5

13. Please indicate your level of agreement with the following statement:

    "The Government should provide a broad umbrella insurance plan that limits liability - similar to the one provided for the nuclear power industry - for the environmental market."

Strongly Disagree	Disagree	Agree	Strongly Agree
1	2	3	4

C. Technology Issues

14. Is technology in the environmental market primarily driven by regulation or market forces?

    Regulation ___   Market ___

15. Is there research and development of environmental technology in your country (meaning the country of your company's headquarters)?

    Yes ___   No ___

    If Yes, how important has the government been in funding research and development of environmental technology in your country?

Not Important		Important		Very Important
1	2	3	4	5

16. For your firm, how important is the development of proprietary technology versus using off-the-shelf technology for the environmental market?

Not Important		Important		Very Important
1	2	3	4	5

SURVEY QUESTIONNAIRE: CONTRACTORS, DESIGNERS, AND TECHNOLOGY VENDORS

17. Has potential liability restricted in any way the development and use of new technology in the environmental market?

Not Restricted		Moderately Restricted		Seriously Restricted
1	2	3	4	5

18. For the purposes of this question, preventive technology prevents ecological damage, whereas corrective technology remedies prior damage.

    a. In the past, has environmental technology been directed toward preventive or corrective purposes?

Preventive				Corrective
<---------------------------------->				
1	2	3	4	5

    b. In the future, will environmental technology be directed toward preventive or corrective purposes?

Preventive				Corrective
<---------------------------------->				
1	2	3	4	5

    c. Is there a desired balance between preventive and corrective technologies?

Preventive				Corrective
<---------------------------------->				
1	2	3	4	5

19. Should the development of environmental technology (preventive and corrective) be promoted by the government of your country (meaning the country of your company's headquarters?

    Yes ___   No ___

    If Yes, how should the government promote technology - (1) directly through sponsorship, or (2) indirectly through incentive programs?

Sponsorship				Incentives
<---------------------------------->				
1	2	3	4	5

20. Please indicate the competitiveness of environmental technology from the following regions and/or countries of the world.

	Not Competitive	Moderately Competitive	Very Competitive	Don't Know
a. United States	1	2	3	9
b. Europe	1	2	3	9
c. Japan	1	2	3	9

SURVEY QUESTIONNAIRE: CONTRACTORS, DESIGNERS, AND TECHNOLOGY VENDORS

21. Please indicate the degree of development of environmental technology in your country (meaning the country of your company's headquarters) and worldwide.

	Inadequate				Adequate	Non-existent
	<-------------------------------->					
a. your country	1	2	3	4	5	9
b. worldwide	1	2	3	4	5	9

22. Is the cost of environmental technology a barrier to entry in the following areas?

	Not a Barrier		Modest Barrier		Strong Barrier	Don't Know
a. Petrochemicals	1	2	3	4	5	9
b. Water Supply	1	2	3	4	5	9
c. Sewer / Waste Water	1	2	3	4	5	9
d. Solid Waste	1	2	3	4	5	9
e. Hazardous Waste	1	2	3	4	5	9
f. Energy	1	2	3	4	5	9
g. Pollution Abatement	1	2	3	4	5	9
h. Environmental Assessment	1	2	3	4	5	9

23. In the environmental market, are innovative or alternative technologies generally favored by funding agencies?

Yes ___ No ___

If <u>Yes</u>, how much emphasis is placed on the utilization of these technologies?

Little Emphasis		Moderate Emphasis		Strong Emphasis
1	2	3	4	5

If <u>No</u>, how much emphasis should be placed on innovative and alternative technologies?

No Emphasis		Moderate Emphasis		Strong Emphasis
1	2	3	4	5

SURVEY QUESTIONNAIRE: CONTRACTORS, DESIGNERS, AND TECHNOLOGY VENDORS

Section III: Response to Environmental Market

The previous section was intended to assess your general perceptions of the environmental market. This section explores how your company is responding to this market. The section is divided into two parts. Part A examines if and how your company is adapting its business strategy to meet the needs of the environmental market. Part B investigates the effects of the public's increasing environmental awareness and responsibility on the construction industry.

A. Business Strategy

If your company currently operates in the environmental market, please answer Questions 1-7 and then proceed to part B.

If your company <u>does not</u> currently operate in the environmental market, please skip Questions 1-7 and proceed to Question 8.

---

1. Please check the strategies that your company has used to acquire the capability to provide environmental services, and judge the importance of each.

		Slightly Important		Moderately Important		Very Important
a. mergers and acquisitions	___	1	2	3	4	5
b. joint ventures	___	1	2	3	4	5
c. established in-house capability	___	1	2	3	4	5

2. At present, what percentage of your work is in the environmental market?
   a. <1% ___ b. 1-10% ___ c. 11-20% ___ d. 21-30% ___
   e. 31-40% ___ f. 41-50% ___ g. 51-60% ___ h. 61-70% ___
   i. 71-80% ___ j. 81-90% ___ k. >90% ___

3. Compared to the <u>present,</u> how important will environmental markets be for your firm in the next 10 years?

Less important		About the same		More important
1	2	3	4	5

4. Please indicate the reasons for operating in the environmental market:
   a. Size or growth potential of environmental market.   ___
   b. The expected profitability of the market.   ___
   c. Competitive advantage over competitors due to:
      i. low cost   ___
      ii. differentiated service   ___
   d. Others [Please specify: _____]

251

SURVEY QUESTIONNAIRE: CONTRACTORS, DESIGNERS, AND TECHNOLOGY VENDORS

5. Please indicate which statement applies to your company:

   a. The company formed exclusively to provide environmental services. ____
   b. The company expanded operations to include environmental services. ____

6. Please indicate the approximate annual rate of growth for your company in the environmental area over the past 5 years in terms of (i) employees and (ii) revenue.

   i. annual growth in number of employees in environmental area
      a. <1% ____   b. 1-5% ____   c. 5-10% ____   d. 11-20% ____
      e. 21-30% ____   f. 31-40% ____   g. >40% ____

   ii. annual growth in revenue from the environmental area
      a. <1% ____   b. 1-5% ____   c. 6-10% ____   d. 11-20% ____
      e. 21-30% ____   f. 31-40% ____   g. >40% ____

7. Please indicate how long your firm has operated in the following segments of the environmental market, and indicate any segments you are planning to enter.

	<1yr	1-5yrs	6-10yrs	>10yrs	Plan to Enter	Do not Plan to Enter
a. Petrochemicals	1	2	3	4	5	0
b. Water Supply	1	2	3	4	5	0
c. Sewer / Waste Water	1	2	3	4	5	0
d. Solid Waste	1	2	3	4	5	0
e. Hazardous Waste	1	2	3	4	5	0
f. Energy	1	2	3	4	5	0
g. Pollution Abatement	1	2	3	4	5	0
h. Environmental Assessment	1	2	3	4	5	0

SURVEY QUESTIONNAIRE: CONTRACTORS, DESIGNERS, AND TECHNOLOGY VENDORS

Please answer Questions 8-11 only if your company does not currently operate in the environmental market.

---

8. Have you considered entering the environmental market and subsequently chosen not to enter?

   Yes ___ No ___

   If Yes, for what reason(s) have you decided against operating in this market?
   a. low profitability ___
   b. potential liability ___
   c. high cost of insurance ___
   d. difficulty in obtaining bonding ___
   e. high cost of technology and equipment ___
   f. other [please specify: _____]

9. Please indicate which statement applies to your company:
   a. The company is planning to provide environmental services. ___
   b. The company has no plans to provide environmental services. ___

10. How likely is it that your firm will consider entering the environmental market in near future?

Not Likely		Likely		Very Likely
1	2	3	4	5

11. Does your company use environmental services provided by other suppliers?

    Yes ___ No ___

    If Yes, rate the relative importance of such services for your operation:

Not Important		Moderately Important		Very Important
1	2	3	4	5

SURVEY QUESTIONNAIRE: CONTRACTORS, DESIGNERS, AND TECHNOLOGY VENDORS

B. Environmental Awareness and Responsibility

Environmental awareness and responsibility have increasingly gained worldwide attention in recent years. This environmental movement has resulted in greater scrutiny of many human activities and their impacts on our environment. New environmental legislation and regulation have been enacted to curb activities deemed deleterious to the environment. As an industry which designs and constructs the built environment, the construction industry would seem particularly prone to intense scrutiny of its actions. Construction necessarily involves changing the environment in which we live and work. Will engineering, design, and construction firms strive to become "green" companies? How important is it for these companies to be perceived as "friends of the environment"? This section is intended to assess the implications of this increased environmental awareness and responsibility on the construction industry.

Two aspects will be considered when talking of "greening" construction companies: (1) the environmental impact of day-to-day operations of the firm, or how it typically conducts business, and (2) the environmental impact of the projects which it undertakes.

1. In general, is the construction industry in your country concerned about the environmental impact of its activities?

Not Concerned		Moderately Concerned		Very Concerned
1	2	3	4	5

2. Has the construction industry received public scrutiny (e.g., from environmental organizations) regarding the environmental impact of its activities?

No Scrutiny		Moderate Scrutiny		Much Scrutiny
1	2	3	4	5

3. Has the construction industry altered its practices due to environmental concerns?

Not Altered		Moderately Altered		Heavily Altered
1	2	3	4	5

4. Please indicate your level of agreement with the following statement:

    "The construction industry is likely to be <u>increasingly</u> scrutinized with regards to the impacts of its activities on the environment."

Strongly Disagree	Disagree	Agree	Strongly Agree
1	2	3	4

SURVEY QUESTIONNAIRE: CONTRACTORS, DESIGNERS, AND TECHNOLOGY VENDORS

5. In the future, what level of <u>public scrutiny</u> will be directed at the construction industry regarding the environmental impacts of its activities?

Less	Same	Little More	More	Much More
1	2	3	4	5

6. In the future, what level of <u>self-scrutiny</u> will the construction industry conduct regarding the environmental impacts of its activities?

Less	Same	Little More	More	Much More
1	2	3	4	5

7. Is it important for the construction industry to be environmentally responsible:

	Not Important		Important		Very Important
a. for the purposes of obtaining work?	1	2	3	4	5
b. for public relations?	1	2	3	4	5

8. Please indicate your level of agreement with the following statement:

   "Engineering, design, and construction firms should actively try to promote themselves as stewards and protectors of the environment."

Strongly Disagree	Disagree	Agree	Strongly Agree
1	2	3	4

9. Will construction firms (engineers, designers, and constructors) be able to create a significant market for themselves as "green" firms? In other words, will construction firms that demonstrate a commitment to environmental awareness and responsibility realize a competitive advantage over those construction firms which do not express similar environmental concerns?

   Yes ___   No ___

   If <u>Yes</u>, will this "green" image be important with regards to obtaining future:

   a. <u>public</u> construction contracts?

Not Important		Moderately Important		Very Important
1	2	3	4	5

   b. <u>private</u> construction contracts?

Not Important		Moderately Important		Very Important
1	2	3	4	5

SURVEY QUESTIONNAIRE: CONTRACTORS, DESIGNERS, AND TECHNOLOGY VENDORS

10. Have new construction projects been restricted in recent years due to environmental concerns?

Not Restricted		Moderately Restricted		Very Restricted
1	2	3	4	5

11. Compared to now, how will future construction projects be restricted due to environmental concerns?

Less Restricted	Same	More Restricted
1	2	3

    If <u>More</u> Restricted, to what degree?

Slightly		Moderately		Much More
1	2	3	4	5

12. Please indicate your level of agreement with the following statements:

    a. "Construction companies should decline to bid or undertake a project which - according to widespread public <u>and</u> scientific opinion - is believed to be detrimental to the environment."

Strongly Disagree	Disagree	Agree	Strongly Agree
1	2	3	4

    b. "The construction industry will <u>not</u> suffer as a consequence of increased public pressure due to the environmental concerns of its actions."

Strongly Disagree	Disagree	Agree	Strongly Agree
1	2	3	4

END.

THANK YOU.

APPENDIX C
SURVEY QUESTIONNAIRE: OWNERS OF CONSTRUCTED FACILITIES

SURVEY QUESTIONNAIRE: OWNERS OF CONSTRUCTED FACILITIES

July 31, 1991

Dear Sir / Madam,

I am conducting a survey of the current environmental construction market, and the enclosed questionnaire will be the foundation of my research. The questionnaire is being sent to 450 companies that operate directly or indirectly in the construction industry. The selected types of companies include: designers, constructors, suppliers, and clients. The questionnaire is designed to investigate several aspects of the environmental construction market - What are the opportunities? What are the challenges? and How is the industry responding to these opportunities and challenges?

The results of the survey will provide the basis for my master's thesis, which is a requirement for a master's degree in civil engineering. I realize that you are often asked to complete various questionnaires and surveys. With this in mind, I have strived to make this one clear and succinct. I need your help to make this questionnaire successful; please complete and return it as soon as possible.

The Center for Construction Research and Education (CCRE) at the Massachusetts Institute of Technology (MIT) is examining the impacts of global environmental change on the construction industry. If you are interested in finding out more about the Center for Construction Research and Education at MIT, or about our research in the environmental area specifically, please feel free to contact us. The Center for Construction Research and Education has traditionally had strong relations with industry and we feel this contact significantly adds to the educational experience at MIT. While much of our contact with industry is direct (e.g., symposiums, guest lectures, courses taught by industry professionals), an important avenue of communication with industry is through surveys and questionnaires such as this one. Once compiled, I will gladly share the findings of the questionnaire with you. We value your support and look forward to hearing from you soon.

Sincerely,

Edmund S. Pendleton

SURVEY QUESTIONNAIRE: CONTRACTORS, DESIGNERS, AND TECHNOLOGY VENDORS

General Information about the Questionnaire

The questionnaire is divided into three sections. Section I asks for general information about your firm, Section II examines the environmental market, and Section III inquires about the responses of your company to this market.

You have been selected to complete the questionnaire because of your firm's involvement in the construction industry. It is important that I know the position or title of the individual that completes the questionnaire. This information will be confidential, but will be needed to classify your response and for any future correspondence. Your answers to the questions will be kept confidential and will only be used in statistical forms. If you feel uncomfortable answering a particular question, please strike through the question number and continue with the next question. However, the validity of the questionnaire is diminished if many questions are omitted.

All of the questions are multiple choice answers except for several fill-in-the-blank questions in Section I. For the multiple choice questions, you are typically asked to respond by: (1) marking the given blank beside your choice, or (2) circling a given number in a graduated scale. I ask that you answer the questions from your own personal viewpoint, not based on company policy or public opinion.

Please complete and return the questionnaire as soon as possible. If you have any questions, please call me at the phone numbers listed below. Thank you for your cooperation.

EDMUND S. PENDLETON
M.I.T., Department of Civil Engineering
77 Massachusetts Ave.
Room 1-050
Cambridge, MA 02139

Office: (617) 253-9736
Home: (617) 266-1412

When returned, this page will be immediately <u>removed</u> from the questionnaire in order to protect the anonymity of your responses.

The following information will <u>only</u> be used to track which companies and individuals have returned the questionnaires, and for any future correspondence.

Optional questions.

1. Please indicate your name and title below.

   Name _____

   Title _____

2. The name of your company is:

   _____

SURVEY QUESTIONNAIRE: OWNERS OF CONSTRUCTED FACILITIES

If returned, this page will be immediately removed from the questionnaire in order to protect the anonymity of your responses.

As a follow up to this questionnaire, I hope to further investigate the environmental construction market by interviewing representatives from several firms in the market. If you would be willing to participate in an interview - which will basically cover the same information in this questionnaire, but in more depth - please complete the following form and return it with the questionnaire.

Name: _____

Title: _____

Company: _____

Address: _____

_____

_____

Phone number: _____

Best times to call: _____

SURVEY QUESTIONNAIRE: OWNERS OF CONSTRUCTED FACILITIES

SURVEY QUESTIONNAIRE (OWNER)
Section I: General Information about Your Firm

The following questions are intended to provide general information about your firm and the types of work that you perform. This general information will be used to categorize your company within the total survey sample, thus allowing comparisons and contrasts with other firms. The specific questions asking for your name and company name are optional. This information would be very helpful for tracking your response to this questionnaire, but it will not be used to identify you, your company, or your responses to any of the questions.

1. When was your company formed?

   a. before 1950 ___   b. 1950 - 1969 ___   c. 1970 - 1979 ___
   d. 1980 - 1989 ___   e. after 1989 ___

2. What is the size of your company?

   a. annual gross revenue* _____

   *Note: Please indicate if you report a figure besides annual gross revenue. For example, many design firms report total annual billings as a measure of financial size.

   b. number of employees: 1-10 ___   11-100 ___   101-1000 ___   >1000 ___

3. For the purposes of the survey, it is necessary to determine your company's role in the construction industry. You may be a client or supplier of the industry rather than a constructor or designer. Please use the following definitions to classify your company.

   owner: owns constructed facilities (e.g., oil companies, utilities)
   constructor: undertakes construction projects (i.e., a builder)
   designer: designs construction projects (e.g., architects, engineers)
   technology vendor: develops and/or sells technology for environmental field
   supplier: supplies materials, equipment, etc., for construction

   With regards to the construction industry, please indicate the principal role of your company in its regular business activities. If your company fills several of these roles, please mark all that apply.

   a. public owner ___
   b. private owner ___
   c. engineering or design firm ___
   d. technology vendor ___
   e. supplier ___
   f. constructor: general contractor ___   specialty (sub) contractor ___

SURVEY QUESTIONNAIRE: OWNERS OF CONSTRUCTED FACILITIES

4. Is your company a subsidiary of another company?

   Yes ___   No ___

5. Please indicate the extent of your company's involvement - measured as a percentage of total business activity - in the following areas.

   *Note: Not at all = 0%   Minor = < 10%   Modest = 11-30%
            Major = 30-99 %   Exclusive = 100%

	Not at all	Minor	Modest	Major	Exclusive
a. Petrochemicals	1	2	3	4	5
b. Water Supply	1	2	3	4	5
c. Sewer / Waste Water	1	2	3	4	5
d. Solid Waste	1	2	3	4	5
e. Hazardous Waste	1	2	3	4	5
f. Energy	1	2	3	4	5
g. Pollution Abatement	1	2	3	4	5
h. Environmental Assessment	1	2	3	4	5

6. Your company headquarters is located in what country?

   _____

7. Please indicate the geographical scope of your company by giving the percentage of your work which is:

   a. local ___%   b. regional ___%   c. national ___ %   d. international ___%

8. Please indicate the extent of your company's operations in each of the listed global regions. (Measured as a percentage of total business operations.)

   Note: Not at all = 0%   Minor = < 10%   Modest = 11-30%
           Major = 30-99 %   Exclusive = 100%

	Not at all	Minor	Modest	Major	Exclusive
a. North America	1	2	3	4	5
b. West Europe	1	2	3	4	5
c. East Europe	1	2	3	4	5
d. Latin America	1	2	3	4	5
e. Far-East	1	2	3	4	5
f. Mid-East	1	2	3	4	5
g. Other _____	1	2	3	4	5

SURVEY QUESTIONNAIRE: OWNERS OF CONSTRUCTED FACILITIES

9. Please indicate the frequency of the following contracting methods for construction work contracted by your firm?

	Never	Not Often	Often	Very Often	Always
a. competitive bidding	1	2	3	4	5
b. cost plus fee	1	2	3	4	5
c. negotiated price	1	2	3	4	5
d. other [please specify: _____]	1	2	3	4	5

SURVEY QUESTIONNAIRE: OWNERS OF CONSTRUCTED FACILITIES

Section II: The Environmental Market

Many dimensions of the rapidly emerging environmental market offer promise for the engineering and construction industry. For the purposes of this survey, the environmental market will include business opportunities derived from the following areas of global environmental concern:

> Acid Rain / Ozone Depletion / Global Warming, Clean Air, Deforestation, Energy, Solid Waste Management, Water Supply, Sewage / Waste Water Treatment, Environmental Assessment Technologies, and Hazardous Waste Remediation and Management.

This section is identical to Section II from a previous survey sent to designers and constructors. While some of these questions were initially intended for designers and constructors, I would be very interested to see your responses - as an owner of constructed facilities - to these same questions. If you are unable to answer a question please strike through it and proceed with the next one. However, I would be grateful if you answer as many questions as possible. This section serves as an introduction into Section III, which contains questions specifically designed for owners.

A. Market Size, Location, and Type

1. In your estimate, what is the approximate <u>annual</u> size (in current US dollars) of the total environmental market (defined above in the introduction to Section II):

    i. in your country (meaning the country of your company's headquarters) today?

        a. < 1 billion ___   b. 1-10 billion ___   c. 11-50 billion ___

        d. 51-100 billion ___   e. > 100 billion ___

    ii. worldwide today?

        a. < 1 billion ___   b. 1-10 billion ___   c. 11-50 billion ___

        d. 51-100 billion ___   e. > 100 billion ___

    iii. in your country (meaning the country of your company's headquarters) in the year 2000?

        a. < 1 billion ___   b. 1-10 billion ___   c. 11-50 billion ___

        d. 51-100 billion ___   e. > 100 billion ___

    iv. worldwide in the year 2000?

        a. < 1 billion ___   b. 1-10 billion ___   c. 11-50 billion ___

        d. 51-100 billion ___   e. > 100 billion ___

SURVEY QUESTIONNAIRE: OWNERS OF CONSTRUCTED FACILITIES

2. Between now and the year 2000 do you expect the environmental market to:
   a. shrink ____   b. stabilize ____   c. grow ____

   If you expect it to <u>grow</u>, at what rate:

Slowly		Moderately		Very Quickly
1	2	3	4	5

3. Please indicate the relative (compared to each other) <u>current</u> environmental market sizes in these regions of the world.

		Small		Medium		Large
a.	North America	1	2	3	4	5
b.	West Europe	1	2	3	4	5
c.	East Europe	1	2	3	4	5
d.	Latin America	1	2	3	4	5
e.	Far-East	1	2	3	4	5
f.	Mid-East	1	2	3	4	5
g.	Other	1	2	3	4	5

4. Please indicate the potential opportunity over the next five years - in the country of your company's headquarters - for the construction industry in the following areas:

		Minor		Modest		Major	Don't Know
a.	Petrochemicals	1	2	3	4	5	9
b.	Water Supply	1	2	3	4	5	9
c.	Sewer / Waste Water	1	2	3	4	5	9
d.	Solid Waste	1	2	3	4	5	9
e.	Hazardous Waste	1	2	3	4	5	9
f.	Energy	1	2	3	4	5	9
g.	Pollution Abatement	1	2	3	4	5	9
h.	Environmental Assessment	1	2	3	4	5	9

SURVEY QUESTIONNAIRE: OWNERS OF CONSTRUCTED FACILITIES

B. Regulatory and Legal Issues

5. How would you rate the current environmental regulation in the following areas?

	Excessive	Not Excessive	Don't Know
a. Petrochemicals	___	___	___
b. Water Supply	___	___	___
c. Sewer / Waste Water	___	___	___
d. Solid Waste	___	___	___
e. Hazardous Waste	___	___	___
f. Energy	___	___	___
g. Pollution Abatement	___	___	___
h. Environmental Assessment	___	___	___

6. To what degree should the environmental market be driven by regulation versus market forces?

100% Reg.	75% Reg.	50% Reg.	25% Reg.	0% Reg.
0% Mar.	25% Mar	50% Mar.	75% Mar.	100% Mar.
1	2	3	4	5	6	7	8	9

7. To what degree should new regulations be punitive versus incentive based?

100% Pun.	75% Pun.	50% Pun.	25% Pun.	0% Pun.
0% Incen.	25% Incen.	50% Incen.	75% Incen.	100% Incen.
1	2	3	4	5	6	7	8	9

8. How important will the following levels of government be in determining the environmental regulations of the future?

	Not Important		Moderately Important		Very Important
a. local	1	2	3	4	5
b. state	1	2	3	4	5
c. federal	1	2	3	4	5

9. a. With regard to <u>environmental</u> regulations, how important is it to establish international regulations?

Not Important		Moderately Important		Very Important
1	2	3	4	5

b. In the absence of international regulations, how important is it for neighboring nations to have comparable environmental regulations?

Not Important		Moderately Important		Very Important
1	2	3	4	5

10. a. What is the likelihood that future environmental regulations will result in additional work for the construction industry?

Low Probability		Moderate Probability		High Probability
1	2	3	4	5

b. What is the likelihood that future environmental regulations will result in greater restrictions on construction?

Low Probability		Moderate Probability		High Probability
1	2	3	4	5

11. In your opinion, is potential liability from environmental work a significant hindrance to operating in the environmental market?

Not Significant		Significant		Very Significant
1	2	3	4	5

12. When compared to other construction markets, to what degree are the following institutions wary of the environmental market because of potential liability issues?

	Less Wary		Same		More Wary
a. lending companies	1	2	3	4	5
b. insurance companies	1	2	3	4	5
c. bonding companies	1	2	3	4	5

13. Please indicate your level of agreement with the following statement:

"The Government should provide a broad umbrella insurance plan that limits liability - similar to the one provided for the nuclear power industry - for the environmental market."

Strongly Disagree	Disagree	Agree	Strongly Agree
1	2	3	4

SURVEY QUESTIONNAIRE: OWNERS OF CONSTRUCTED FACILITIES

C. Technology Issues

14. Is technology in the environmental market primarily driven by regulation or market forces?

    Regulation ___   Market ___

15. Is there research and development of environmental technology in your country (meaning the country of your company's headquarters)?

    Yes ___   No ___

    If <u>Yes</u>, how important has the government been in funding research and development of environmental technology in your country?

Not Important		Important		Very Important
1	2	3	4	5

16. For your firm, how important is the development of proprietary technology versus using off-the-shelf technology for the environmental market?

Not Important		Important		Very Important
1	2	3	4	5

17. Has potential liability restricted in any way the development and use of new technology in the environmental market?

Not Restricted		Moderately Restricted		Seriously Restricted
1	2	3	4	5

18. For the purposes of this question, preventive technology prevents ecological damage, whereas corrective technology remedies prior damage.

    a. In the past, has environmental technology been directed toward preventive or corrective purposes?

Preventive				Corrective
<---------------------------------->				
1	2	3	4	5

    b. In the future, will environmental technology be directed toward preventive or corrective purposes?

Preventive				Corrective
<---------------------------------->				
1	2	3	4	5

SURVEY QUESTIONNAIRE: OWNERS OF CONSTRUCTED FACILITIES

c. Is there a desired balance between preventive and corrective technologies?

```
        Preventive                    Corrective
        <----------------------------------->
            1       2       3       4       5
```

19. Should the development of environmental technology (preventive and corrective) be promoted by the government of your country (meaning the country of your company's headquarters)?

    Yes ___   No ___

    If <u>Yes</u>, how should the government promote technology - (1) directly through sponsorship, or (2) indirectly through incentive programs?

```
        Sponsorship                    Incentives
        <----------------------------------->
            1       2       3       4       5
```

20. Please indicate the competitiveness of environmental technology from the following regions and/or countries of the world.

	Not Competitive	Moderately Competitive	Very Competitive	Don't Know
a. United States	1	2	3	9
b. Europe	1	2	3	9
c. Japan	1	2	3	9

21. Please indicate the degree of development of environmental technology in your country (meaning the country of your company's headquarters) and worldwide.

	Inadequate			Adequate		Non-existent
	<-------			------->		
a. your country	1	2	3	4	5	9
b. worldwide	1	2	3	4	5	9

22. Is the cost of environmental technology a barrier to entry in the following areas?

	Not a Barrier		Modest Barrier		Strong Barrier	Don't Know
a. Petrochemicals	1	2	3	4	5	9
b. Water Supply	1	2	3	4	5	9
c. Sewer / Waste Water	1	2	3	4	5	9
d. Solid Waste	1	2	3	4	5	9
e. Hazardous Waste	1	2	3	4	5	9
f. Energy	1	2	3	4	5	9
g. Pollution Abatement	1	2	3	4	5	9
h. Environmental Assessment	1	2	3	4	5	9

SURVEY QUESTIONNAIRE: OWNERS OF CONSTRUCTED FACILITIES

Section III: Response to Environmental Market

The previous section was intended to assess your general perceptions of the environmental market. This section explores how your company - as an owner of constructed facilities - is responding to environmental business issues. The section is divided into two parts. Part A examines if and how your company is adapting its business strategy to accommodate or serve growing environmental concerns. Part B investigates the effects of the public's increasing environmental awareness and responsibility on your construction decisions.

A. Business Strategy

1. Please indicate how long your firm has operated in the following segments of the environmental market, and indicate any segments you are planning to enter.

	< 1yr	1-5yrs	6-10yrs	>10yrs	Plan to Enter	Do not Plan to Enter
a. Petrochemicals	1	2	3	4	5	0
b. Water Supply	1	2	3	4	5	0
c. Sewer / Waste Water	1	2	3	4	5	0
d. Solid Waste	1	2	3	4	5	0
e. Hazardous Waste	1	2	3	4	5	0
f. Energy	1	2	3	4	5	0
g. Pollution Abatement	1	2	3	4	5	0
h. Environmental Assessment	1	2	3	4	5	0

2. Compared to the <u>present,</u> how important will these segments of the environmental market be for your firm in the next 10 years?

	Less important		About the same		More important
a. Petrochemicals	1	2	3	4	5
b. Water Supply	1	2	3	4	5
c. Sewer / Waste Water	1	2	3	4	5
d. Solid Waste	1	2	3	4	5
e. Hazardous Waste	1	2	3	4	5
f. Energy	1	2	3	4	5
g. Pollution Abatement	1	2	3	4	5
h. Environmental Assessment	1	2	3	4	5

SURVEY QUESTIONNAIRE: OWNERS OF CONSTRUCTED FACILITIES

3. Have environmental regulations and concerns shaped the direction of your industry in recent years?

Not at all	Slightly	Moderately	Significantly	Very Significantly
1	2	3	4	5

4. Have the following areas of your business been affected by environmental concerns and regulations?

	Not at all	Slightly	Moderately	Significantly	Very Significantly
a. resource or materials input	1	2	3	4	5
b. production / process	1	2	3	4	5
c. distribution	1	2	3	4	5
d. marketing	1	2	3	4	5
e. service	1	2	3	4	5

5. Do you expect a need for new constructed facilities (in your industry) <u>due to</u> environmental regulations and concerns?

   A list of facility categories is given below. For example, installing smokestack scrubbers to meet new clean air regulations would fall under category c: air pollution abatement. Please feel free to add other categories.

   a. new or modified production/process facilities _____
   b. energy supply facilities _____
   c. air pollution abatement facilities _____
   d. treatment or containment facilities for:
      solid waste _____   waste water _____   hazardous waste _____
   e. waste reduction reduction facilities for:
      solid waste _____   waste water _____   hazardous waste _____
   f. other: _____

SURVEY QUESTIONNAIRE: OWNERS OF CONSTRUCTED FACILITIES

Note:

For the purposes of this survey, environmental services are those services which meet needs derived from the following areas of global environmental concern:

> Acid Rain / Ozone Depletion / Global Warming, Clean Air, Deforestation, Energy, Solid Waste Management, Water Supply, Sewage / Waste Water Treatment, Environmental Assessment Technologies, and Hazardous Waste Remediation and Management.

---

6. Have recent environmental regulations, restrictions, and awareness increased your need for environmental services?

    Yes ____   No ____

7. For your industry, how do you foresee the need for environmental services between now and the year 2000?

Much Less	Less	Same	More	Much More
1	2	3	4	5

8. Please check the strategies that your company has used to acquire environmental services, and judge the importance of each.

		Slightly Important		Moderately Important		Very Important
a. mergers and acquisitions	____	1	2	3	4	5
b. joint ventures	____	1	2	3	4	5
c. established in-house capability	____	1	2	3	4	5
d. hired outside supplier	____	1	2	3	4	5
e. no services needed	____					

9. Does your company <u>use</u> environmental services provided by other suppliers?

    Yes ____   No ____

    If <u>Yes</u>, rate the relative importance of such services for your operation:

Not Important		Moderately Important		Very Important
1	2	3	4	5

SURVEY QUESTIONNAIRE: OWNERS OF CONSTRUCTED FACILITIES

10. Have you considered establishing environmental services in-house and subsequently chosen not to?

    Yes ___ No ___

    If Yes, for what reason(s) have you decided against establishing these services?
    a. low profitability ___
    b. potential liability ___
    c. high cost of insurance ___
    d. difficulty in obtaining bonding ___
    e. high cost of technology and equipment ___
    f. other [please specify: _____]

SURVEY QUESTIONNAIRE: OWNERS OF CONSTRUCTED FACILITIES

B. Environmental Awareness and Responsibility

Environmental awareness and responsibility have increasingly gained worldwide attention in recent years. This environmental movement has resulted in greater scrutiny of many human activities and their impacts on our environment. New environmental legislation and regulation have been enacted to curb activities deemed deleterious to the environment. As an industry which designs and constructs the built environment, the construction industry would seem particularly prone to intense scrutiny of its actions. Construction necessarily involves changing the environment in which we live and work. As owners in the construction industry strive to become "green" companies, how will construction decisions be affected? This section is intended to assess the implications of increased environmental awareness and responsibility on your construction decisions as an owner. In addition, several questions refer to the construction industry in general.

1. In general, are owners of constructed facilities (in your country) concerned about the environmental impact of their construction activities?

Not Concerned		Moderately Concerned		Very Concerned
1	2	3	4	5

2. Have owners received public scrutiny (e.g., from environmental organizations) regarding the environmental impact of their <u>construction</u> activities?

No Scrutiny		Moderate Scrutiny		Much Scrutiny
1	2	3	4	5

3. Have owners of constructed facilities altered their construction decisions due to environmental concerns?

Not Altered		Moderately Altered		Heavily Altered
1	2	3	4	5

4. Please indicate the appropriate answers to the following questions:

    "Is it important for your business to be environmentally responsible:

	Not Important		Important		Very Important
a. for selling products or services?"	1	2	3	4	5
b. for public relations?"	1	2	3	4	5

SURVEY QUESTIONNAIRE: OWNERS OF CONSTRUCTED FACILITIES

5. In the future, what level of <u>public scrutiny</u> will be directed at owners regarding the environmental impacts of their construction activities?

Less	Same	Little More	More	Much More
1	2	3	4	5

6. In the future, what level of <u>self-scrutiny</u> will owners conduct regarding the environmental impacts of their construction activities?

Less	Same	Little More	More	Much More
1	2	3	4	5

7. Please indicate your level of agreement with the following statement:

   "Owners of constructed facilities should actively try to promote themselves as stewards and protectors of the environment."

Strongly Disagree	Disagree	Agree	Strongly Agree
1	2	3	4

8. In your opinion as an owner, will construction firms (engineers, designers, and constructors) be able to create a significant market for themselves as "green" firms? In other words, will construction firms that demonstrate a commitment to environmental awareness and responsibility realize a competitive advantage over those construction firms which do not express similar environmental concerns?

   Yes ___   No ___

   If <u>Yes</u>, will this "green" image be important with regards to obtaining future:

   a. <u>public</u> construction contracts?

Not Important		Moderately Important		Very Important
1	2	3	4	5

   b. <u>private</u> construction contracts?

Not Important		Moderately Important		Very Important
1	2	3	4	5

9. As an owner, would you be more likely to hire a construction firm (engineer, designer, constructor) that has an established record of commitment to the environment versus one that does not? (Assume the two firms are comparable in other regards.)

   Yes ___   No ___

SURVEY QUESTIONNAIRE: OWNERS OF CONSTRUCTED FACILITIES

10. In recent years, have new construction projects in your industry been restricted due to environmental concerns?

Not Restricted		Moderately Restricted		Very Restricted
1	2	3	4	5

11. Compared to now, how will future construction projects in your industry be restricted due to environmental concerns?

Less Restricted	Same	More Restricted
1	2	3

    If <u>More</u> Restricted, to what degree?

Slightly		Moderately		Much More
1	2	3	4	5

END.
THANK YOU.

SURVEY QUESTIONNAIRE: OWNERS OF CONSTRUCTED FACILITIES

January 10, 1992

Dear Sir / Madam,

I am conducting a survey of the current environmental construction market, and the enclosed questionnaire will be the foundation of my research. The questionnaire is being sent to 450 companies that operate directly or indirectly in the construction industry. The selected types of companies include: designers, constructors, suppliers, and owners of constructed facilities. The questionnaire is designed to investigate several aspects of the environmental construction market - What are the opportunities? What are the challenges? and How is the industry responding to these opportunities and challenges?

This version of the questionnaire has been modified slightly to address issues facing <u>owners</u> of constructed facilities. Sections I and II remain the same, but Section III has been modified to assess how current environmental concerns affect owner's decisions to undertake new construction projects.

The results of the survey will provide the basis for my master's thesis, which is a requirement for a master's degree in civil engineering. I realize that you are often asked to complete various questionnaires and surveys. With this in mind, I have strived to make this one clear and succinct. I need your help to make this questionnaire successful; please complete and return it as soon as possible.

The Center for Construction Research and Education (CCRE) at the Massachusetts Institute of Technology (MIT) is examining the impacts of global environmental change on the construction industry. If you are interested in finding out more about the Center for Construction Research and Education at MIT, or about our research in the environmental area specifically, please feel free to contact us. The Center for Construction Research and Education has traditionally had strong relations with industry and we feel this contact significantly adds to the educational experience at MIT. While much of our contact with industry is direct (e.g., symposiums, guest lectures, courses taught by industry professionals), an important avenue of communication with industry is through surveys and questionnaires such as this one. Once compiled, I will gladly share the findings of the questionnaire with you. We value your support and look forward to hearing from you soon.

Sincerely,

Edmund S. Pendleton

SURVEY QUESTIONNAIRE: OWNERS OF CONSTRUCTED FACILITIES

General Information about the Questionnaire

The questionnaire is divided into three sections. Section I asks for general information about your firm, Section II examines the environmental market, and Section III inquires about your company's responses to construction needs and concerns in this market.

You have been selected to complete the questionnaire because of your firm's involvement in the construction industry. It is important that I know the position or title of the individual that completes the questionnaire. This information will be confidential, but will be needed to classify your response and for any future correspondence. Your answers to all questions will be kept confidential and will only be used in statistical forms. If you feel uncomfortable answering a particular question, please strike through the question number and continue with the next question. However, the validity of the questionnaire is diminished if many questions are omitted.

All of the questions are multiple choice answers except for several fill-in-the-blank questions in Section I. For the multiple choice questions, you are typically asked to respond by: (1) marking the given blank beside your choice, or (2) circling a given number in a graduated scale. I ask that you answer the questions from your own personal viewpoint, not based on company policy or public opinion.

Please complete and return the questionnaire as soon as possible. If you have any questions, please call me at the phone numbers listed below. Thank you for your cooperation.

EDMUND S. PENDLETON
M.I.T., Department of Civil Engineering
77 Massachusetts Ave.   Office: (617) 253-9736
Room 1-050   Home: (617) 266-1412
Cambridge, MA 02139

When returned, this page will be immediately <u>removed</u> from the questionnaire in order to protect the anonymity of your responses.

The following information will <u>only</u> be used to track which companies and individuals have returned the questionnaires, and for any future correspondence.

Optional questions.

1. Please indicate your name and title below.

    Name _____

    Title _____

2. The name of your company is:

    _____

SURVEY QUESTIONNAIRE: OWNERS OF CONSTRUCTED FACILITIES

If returned, this page will be immediately removed from the questionnaire in order to protect the anonymity of your responses.

As a follow up to this questionnaire, I hope to further investigate the environmental construction market by interviewing representatives from several firms in the market. If you would be willing to participate in an interview - which will basically cover the same information in this questionnaire, but in more depth - please complete the following form and return it with the questionnaire.

Name: _____

Title: _____

Company: _____

Address: _____

_____

_____

Phone number: _____

Best times to call: _____

REFERENCES

Introduction

Development and the Environment, World Development Report 1992, The World Bank, New York: Oxford University Press, 1992.

Chapter 1. Environment and the Construction Industry

Construction Opportunities in Environmental Markets; (see Preface).
"Elaborate Sting Operation Brings Arrests in Illegal Dumping of Toxic Wastes by Businesses." *New York Times,* May 13, 1992.
Moavenzadeh, Fred. "A Strategic Response to a Changing Engineering and Construction Market." Background paper prepared for the Engineering and Construction Forum, World Economic Forum, MIT, April 1989.

Chapter 2. Industrial Response and Initiatives

Choucri, Nazli. *Report on New Paradigm for Global Environmental Change.* Working paper No. 1 CCRE February 1992.
Jacobson, A. F., "Business Championing the Environment." Paper presented at the 75th Anniversary of the Conference Board, November 5, 1991.
Niemczynowicz, Janusz. "Environmental Impact of Urban Areas—the Need for Paradigm Change." *Water International* 16 (1991).
Solid Waste Management: Decision and Market Dilemmas; (see Preface).
U.S. Congress. Senate. *Project 88—Harnessing Market Forces to Protect Our Environment: Initiatives for the New President.* A Public Policy Study sponsored by Timothy E. Wirth and John Heinz. December 1988.
U.S. Congress. Senate. *Project 88—Round II—Incentives for Action: Designing Market-Based Environmental Strategies.* A Public Policy Study sponsored by Timothy E. Wirth and John Heinz. May 1991.

Chapter 3. Environmental Policy and Regulations

The Hazardous Waste Remediation Market: Innovative Technological Development and the Growing Involvement of the Construction Industry; (see Preface).
"Accords for Nature's Sake." *New York Times,* June 15, 1992, p. A8.
Graedel, Thomas E. and Crutzen, Paul J. "The Changing Atmosphere." *Scientific American* (September 1989).
Hahn, Robert W., and Stavins, Robert N. "Incentive-Based Environmental Regulation: A New Era from an Old Idea?" *Ecology Law Quarterly* 18:1 (1991).
Landy, Marc K., Roberts, Marc J., and Thomas, Stephen R. "Writing the Resource Conser-

vation and Recovery Regulations." In *The Environmental Protection Agency: Asking the Wrong Questions,* New York: Oxford University Press, 1990.

Lindzen, Richard S. "Global Warming: The Origin and Nature of Alleged Scientific Consensus." Lecture Presentation at MIT, March 11, 1992.

"New Studies Predict Profits in Heading off Warming." *The New York Times,* March 17, 1992.

Policy Implications of Greenhouse Warming. Synthesis Panel, Committee on Science, Engineering, and Public Policy; National Academy of Sciences; National Academy of Engineering; Institute of Medicine. Washington D.C.: National Academy Press, 1991.

Project 88 and *Project 88—Round II;* (see Chap. 2 Refs.)

Rathje, William, and Murphy, Cullen. *Rubbish! The Archaeology of Garbage.* New York: HarperCollins, 1992.

Schneider, Stephen H. "The Changing Climate." *Scientific American* (September 1989).

U.S. Congress. Office of Technology Assessment. *Facing America's Trash: What's Next for Municipal Solid Waste?* Washington, D.C.: U.S. Government Printing Office, October 1989, pp. 284, 350.

Wakita, Kazushi. *"Management of Waste Disposed into the Atmosphere."* Working Paper #4, CCRE, MIT. December 1992.

Chapter 4. Technological Challenges and Opportunities

"A Nation's Recycling Law Puts Businesses on the Spot." *New York Times,* July 12, 1992, p. 5, Business Section.

Arrowsmith, Peter D. "The Challenges of Applying Advanced Technologies to Hazardous Waste Remediation." *Construction Business Review* (July/August 1993).

Bierbaum, Rosina, and Friedman, Robert M. "The Road to Reduced Carbon Emissions." Issues in *Science and Technology* (Winter 1991–92).

Darkazalli, Ghaze. "Energy and the Environment in the Twenty-first Century." In *Future Effect and Contribution of Photovoltaic Electricity on Utilities and the Environment.* Cambridge, Mass: MIT Press, 1990.

Goldstein, Nora, and Spencer, Robert. "Solid Waste Composting in the United States." *BioCycle* (November 1990).

Hahn, R. W., and Stavins, R. N. "Incentive Based Environmental Regulation: A New Era from an Old Idea?" *Ecology Law Quarterly,* **18,** 1–42 (1991).

The Hazardous Waste Remediation Market: Innovative Technological Development and the Growing Involvement of the Construction Industry; (see Chap. 3 Refs.).

"How to Throw Things Away." *The Economist;* April 13, 1991.

Kinnear, James W. "Clean Air at a Reasonable Price." Issues in *Science and Technology* (Winter 1991–92).

Portney, Paul. *Public Policies for Environmental Protection.* Washington, D.C.: Resources for the Future, 1990.

Rubbish! The Archaeology of Garbage; Rathje and Murphy, (see Chap. 3 Refs.).

"Sanitary Landfill Costs Estimated." Walsh, James. *Waste Age* (March, 1990).

Teitenberg, T. *Environmental and Natural Resource Economics.* New York: HarperCollins, 1992.

White, David C. "Global Environment and the Electrical Power Industry: A Period of Conflict, Growth, New Supply, and End Use Technology." Department of Electrical Engineering and Computer Science, MIT, 1991.

White, David C., et al., *The New Team: Electricity Without Carbon Dioxide. Technology Review*, MIT (January 1992).

Chapter 5. Environmental Markets

"Economic and Financial Indicators: Environmental Spending." *The Economist* (October 12, 1991).

"ENR Forecast 1992." *Engineering News Record*, January 27, 1992.

Fairweather, Virginia, "The Environment is Good Business in France." *Civil Engineering* (March 1992).

Franklin Associates, Ltd., "Characterization of Municipal Solid Waste in the United States: 1990 Update." Report by the Municipal Solid Waste Task Force, Office of Solid Waste; Washington, D.C., June 1990.

Future Construction Demands in Environmental Markets (see Preface).

Gentry, Bradford S., and Schurtz, James M. *Developments in the European Environmental Market: Opportunities and Constraints*. Morrison & Foerster (October 1991).

Portney, Paul, *Public Policies for Environmental Protection*. Washington, D.C.: Resources for the Future, 1990.

"The Industry Takes Shape." *Environmental Business Journal* (April 1991).

Chapter 6. Shifting Attitudes Create Opportunities for Construction

Future Construction Demands in Environmental Markets and *Construction Opportunities in Environmental Markets* (see Preface).

Chapter 7. Corporate Strategy and Future Trends

Hoffman, Andrew J., "Weighing the Pros and Cons to Hazardous Waste Remediation." Article prepared for *Construction Business Review*, for Construction Research and Education (April 1993).

Jeijell, Mohamed N., and Russell, Jeffrey S., "Surety Bonds for Hazardous and Toxic Wastes: Opportunities and Risks." *Construction Business Review* (July/August 1993).

Liddle, Brant, "Sustainable Development." Working paper prepared for master's thesis, MIT Center for Construction Research and Education, November 1992.

Marini, Robert C., Gates, Stephen R., and Tunnicliffe, Peter W., "A Return to Teamwork: Partnering Approach Facilitates Hazardous Waste Cleanup." *Construction Business Review* (July/August 1993).

"Pollution Minimization Opportunities and Strategies for Construction Firms." Moores, C. W., *CBR* 3:4 (July/August 1993).

Stavins, Robert N., "Clean Profits: Using Economic Incentives to Protect the Environment." *Policy Review* (Spring 1989).

The Challenges of Applying Advanced Technologies to Hazardous Waste Remediation (see Chap. 4, Refs.)

INDEX

Acid rain, 29
Air pollution:
 environmental market opportunities, 156–157
 Environmental Protection Agency (EPA) and, 80, 81, 99–100
 government policy and, 80–81
Air Pollution Control Act of 1962, 65
Alliances. *See* Teamwork approach
Alternative fuels:
 energy utilities, 124–125, 162
 technology and, xxvi
Ambient standards:
 air pollution control, 80
 command-and-control regulations, 66
 water pollution control, 82
Arrowsmith, Peter D., 130, 215, 236
Asset-management perspective, sustainable development and, 51–52
AT&T pollution prevention program, described, 45–48

Baghouse, solid waste technology and, 117
Batteries, lead pollution, 73–74
Best available technology requirements, command-and-control regulations, 67
Bioremediation:
 communications limitations and, 21–22
 hazardous waste technology, 112–113
Bottle deposit laws. *See* Deposit-refund system
Brown, Jerry, 76
Bush, George, 38, 73, 74, 126
Business strategies, 197–201
 industry, 197–199
 owners, 199–201

Capitalism, economic models, traditional, 33–37
Carbon dioxide, electric power production, 159–160, 161
Carson, Rachel L., 29
Cement-based solidification, hazardous waste technology, 109
CERCLA. *See* Superfund program

Chemical extraction, hazardous waste technology, 110
Chemical treatment, hazardous waste technology, 111
Chernobyl nuclear plant disaster, 158
Choucri, Nazli, 226–227, 229
Citizen empowerment, described, 77–78
Clean Air Act of 1963, 48, 80
Clean Air Act of 1970, 48, 80
Clean Air Act of 1977 Amendments, 99
Clean Air Act of 1990 Amendments, 48, 73, 81, 156, 157, 159
Clean Water Act of 1977, 82, 155
Clinton, William, 127, 128
Coal, energy production and, 121, 122
Coalition of Environmentally Responsible Economies (CERES), 5, 147
Cogeneration, energy production and, 123
Collaborative approach. *See* Teamwork approach
Command-and-control regulations. *See also* Government policy; market-based incentives
 corporate strategies, 220–222
 described, 65–68
 hazardous waste technology and, 105–113
 solid waste technology and, 113–120
 technology and, 96, 99–100, 104
Common market. *See* European Community
Communication, technology and, 21–23
Competition:
 construction industry and, 12, 191–192
 economic models, 34, 36
 market-based incentives, 70
Compliance, timetables and, 104
Composting:
 environmental market opportunities, 147, 152, 153
 solid waste technology and, 118–119
Comprehensive Environmental Response, Compensation and Liability Act of 1980 (CERCLA). *See* Superfund program
Conflict resolution, government policy and, 58, 59

INDEX

Conservation:
  energy generation, 49, 76, 120–121, 160
  public opinion and, xx–xxii
Consortia, technology, barriers to, 128–129
Construction industry. *See also* Corporate policies; Corporate strategies
  crisis and, xix–xx
  environmental issues and, xvii–xviii
  environmental markets, 133–171. *See also* Environmental markets
  globalization and, xxvii
  initiatives of, xxv–xxvi
  lesser developed countries and, xxix–xxx
  marketing and, 3–4
  mission of, xvii
  regulatory impact on, 7–9
  structural changes within, 4–7
  sustainable development and, xxvii–xxviii
  teamwork approach in, 11–16
  technology and, xxvi, 96
Construction industry opportunities, 174–207. *See also* Environmental markets: opportunities in
  business strategies, 197–201
    industry, 197–199
    owners, 199–201
  environmental awareness and opportunities, 201–207
    industry, 202–205
    owners, 205–207
  environmental issues and, xviii, 2–3
  industry overview, 178–182
  market perspectives (industry), 183–192
    market size, location, and type, 183–186
    regulatory and legal issues, 186–190
    technology issues, 190–192
  market perspectives (owners), 182–197, 192–197
    market size, location, and type, 192–193
    regulatory and legal issues, 193–195
    technology issues, 195–197
  overview of, 174–175
  owner characteristics, 182
  survey questionnaire, 176–178
    design of, 176–177
    respondent characteristics, 177–178
Consultation and expertise, construction industry opportunities, 174–175
Corporate policies. *See also* Construction industry

AT&T pollution prevention program, 45–48
  energy utilities, 48–50
  environmental movement opposed, 30–31
  European experience, 52–54
  government policy and, 40–42
  3M pollution prevention program, 42–45
Corporate strategies, 210–223
  opportunities, 223
  overview of, 210–211
  risk and liabilities, 212–215
  shift from regulation to market incentives, 220–222
  valuation of environment, 218–219
  vision required, 216–217
  wealth preservation and expansion, 217–218
Cousteau, Jacques, 29
Crisis, environmental issues and, xix–xx

Defense Advanced Research Projects Agency (DARPA), 129
Department of Defense (DOD). *See also* Military
  environmental market growth, 140
  hazardous waste disposal, 22
  risk and liabilities, 213
Department of Energy (DOE):
  environmental market growth, 140
  hazardous waste disposal, 22
  sustainable development technology, barriers to, 20
Deposit-refund system:
  drawbacks of, 74
  market-based incentives and, 101
Developing countries. *See* Lesser developed countries
Drucker, Peter, 18
Dry scrubber, solid waste technology and, 117–118

Earth Summit. *See* United Nations Conference on Environment and Development (UNCED)
Economic factors:
  command-and-control regulations, 67
  economic models, traditional, 33–37
  fines and penalties, 40
  government policy and, 37–40
  market-based incentives, 68–77. *See also* Market-based incentives
  waste-management approach, 10

INDEX

Economic growth:
  corporate strategies, 217–218
  electric power production, 158
  environmental values and, 61
  international accords and, 89
  sustainable development concept and, 64
Effluent standards, command-and-control regulations, 66
Ehrenfeld, John R., 228–229
Electric utilities. *See also* Energy utilities
  environmental markets and, 158–163
  fuels for, 121
  technology and, 121–124
Electro-osmosis, hazardous waste technology, 111
Electrostatic precipitator, solid waste technology and, 117
Emission standards:
  air pollution control, 81
  command-and-control regulations, 66
  electric power production, 159
  solid waste disposal market opportunities and, 151
  solid waste technology and, 117
Energy efficiency:
  market mechanisms and, 75–76
  technology and, 120–124
Energy utilities. *See also* Electric utilities
  alternative fuels, 124–125
  corporate policies, 48–50
  economic growth and, 61
  fuels for, 121
  international accords and, 91–92
  market mechanisms and, 76
  technology and, xxvi
Enforcement:
  air pollution control, 80–81
  increased emphasis on, 8, 40
Environmental Action Plans (EAPs, EC), 87–88
Environmental assessment, environmental market opportunities, 142–143
Environmental Impact Statement (EIS), 77
Environmental issues:
  construction industry and, xvii–xviii, 2–3
  crisis and, xix–xx
  globalization and, xxvii
  government policy and, xxii–xxiv
  historical perspective on, 27–33, 64–65
  lesser developed countries and, xxix–xxx
  public opinion and, xx–xxii, 32
  sustainable development and, xxiv, xxvii–xxviii
  technology and, xxvi
  trade-offs and, 103
  values, economic growth and, 61
Environmental markets, 133–171. *See also* Markets
  construction industry opportunities, 182–197. *See also* Construction industry opportunities
  demand creation, 134–135
  dynamics of, 136–137
  electric power production and, 158–163
  global perspective on, 165–171
    Europe, 165–169
    Pacific Rim, 169–171
  government policy and, 97–99
  growth in, 140–141
  opportunities in, 142–157. *See also* Construction industry opportunities
    air pollution abatement, 156–157
    environmental assessment, 142–143
    hazardous waste, 143–147
    sewer/wastewater, 154–155
    solid waste, 147–153
    waste minimization, 153–154
    water supply, 155–156
  petrochemicals and, 163–164
  structure of, 137–140
Environmental movement:
  government policy and, 62–63
  industry opposition to, 30–31
  origins of, 29–30
  overview of, 61–63
Environmental organizations, membership statistics in, 62
Environmental Protection Agency (EPA). *See also* Superfund program
  air pollution control, 80, 81, 99–100
  communication limitations and, 21–23
  creation of, 30, 65
  enforcement efforts by, 8, 65
  hazardous waste disposal, 5, 84
  historical perspective on, 32
  risk and liabilities, 213, 214
  solid wastes and, 83–84
  technology, barriers to, 19–20, 129–130
  tradable permits, 72
European Community (EC):
  environmental markets, 165–169

287

European Community (*continued*)
　government policy, 85–89
Expertise and consultation, construction industry opportunities, 174–175
Exxon Corporation, fines and penalties, 40

Fabric filter, solid waste technology and, 117
Federal Technology Transfer Act of 1986, 127
Federal Water Pollution Control Act (FWPCA) of 1972, 82, 154, 155
Fee-based waste management, market-based incentives and, 71–72. *See also* Pollution fees/taxes
Finance, historical perspective on, 28
Fines and penalties, increase in, 40
Food, Drug and Cosmetics Act of 1962, 65
Fossil fuels. *See also* Electric utilities; Energy utilities
　energy production and, 121, 159
　social costs of, alternative fuels versus, 162
Freeze separation, hazardous waste technology, 110–111

Gas turbine combined-cycle (GTCC) system, energy production, 121–122
General Accounting Office (GAO), hazardous waste site statistics, 33
Gentry, Bradford S., 168, 169, 239–240
Geothermal energy, alternative fuels, 125, 162
Glassification, hazardous waste technology, 110
Global perspective:
　construction industry and, 6–7, 180, 181, 188–189
　environmental issues and, xxvii
　environmental markets, 165–171
　　Europe, 165–169
　　Pacific Rim, 169–171
Global warming:
　electric power production, 159, 163
　international accords and, 89–92
　ozone depletion, 90–91
Government policy. *See also* Command-and-control regulations; Market-based incentives; State regulation
　areas of, 79–85
　　airborne waste, 80–81
　　solid and hazardous waste, 82–85
　　waterborne waste, 81–82
　citizen empowerment and, 58–59, 77–78
　command-and-control regulations, 65–68

communications limitations and, 22
conflict resolution and, 58
construction industry opportunities
　market perspectives (industry), 186–190
　market perspectives (owners), 193–195
corporate policies and, 40–42
diversity in, 78–79
energy utilities, 48–50
enforcement and, 8, 40
environmental issues and, xxii–xxiv, 2
environmental markets
　demand creation, 135
　dynamics of, 136–137
environmental movement and, 62–63
European experience, 52–54, 85–89
frustration of, 37–40
hazardous waste disposal and, 3–4
historical perspective on, 32–33, 64–65
industry lobbying efforts and, 31, 79
international accords, 89–93
market-based incentives, 68–77
　described, 68–70
　implementation difficulties, 75–77
　program advantages and disadvantages, 70–75
political factors and, 59–60. *See also* Politics
rationale for, 63–64
regulatory impact on construction industry, 7–9
sustainable development and, xxvii–xxviii
technology and, xxvi, 17, 97–99, 103–105, 125–131
Government subsidies. *See* Subsidies
Granular filter, solid waste technology and, 117
Greenhouse effect. *See* Global warming
Groundwater pollution:
　environmental market growth, 140, 156
　public opinion and, 32
　technological detection of, 16

Hazardous Solid Waste Amendment (HSWA), 84
Hazardous waste disposal:
　command-and-control regulations and technology, 105–113
　corporate policy and, 41
　costs of, 13
　environmental market growth, 140
　environmental market opportunities, 143–147
　environmental market structure, 138
　government policy and, 7–8, 82–85, 129–130

industry initiatives and, xxv
market forces and, 3–4
military, 20, 22, 28
public opinion and, 32
sites, statistics on, 32–33, 105, 144
statistics of, 5
Health hazards:
  alliances and, 14
  public opinion and, 32
Heinz, John, 74
Hoffbuhr, Jack, 156
Hoffman, Andrew J., 144
Hollings, Ernest, 128
Hydroelectric power, energy production technology, 123–124

Incineration:
  environmental market opportunities, 147, 148–149, 151–152
  hazardous waste technology, 107, 108
  solid waste technology, 115–118
Innovation. *See also* Technology
  communications limitations and, 21–23
  technology, sustainable development, barriers to, 16–20, 125–131
In situ systems, hazardous waste technology, 106
In-tank systems, hazardous waste technology, 106
Integrated gasification combined-cycle (IGCC) system, energy production, 122
International accords, government policy and, 89–93. *See also* Global perspective
Investment:
  environmental market opportunities, 153
  government policy and, 97–98
  market-based incentives, 70
  sustainable development technology, barriers to, 16–17, 19
Ion exchange, hazardous waste technology, 110

Jacobson, A. F., 42
Jain, Ravi, 230–231
Johnson, Lyndon B., 64, 65

Kagami, Akira, 240
Kovalick, Walter W., Jr., 19, 233–234

Labor relations, alliances and, 14
Landfills:
  closing of, 83
  environmental market opportunities, 147, 148–150
  solid waste technology and, 114–115
Laudise, R. A., 229–230
Law. *See* Legal issues; Liability
Lead pollution, batteries, 73–74
Legal issues. *See also* Liability
  industry market perspectives, 186–190
  owner market perspectives, 193–195
Legislation. *See* Government policy
Lesser developed countries. *See also* Global perspective
  energy production technology, 123–124
  environmental issues and, xxix–xxx
  international accords and, 89
Liability. *See also* Legal issues
  alliances and, 14
  construction industry opportunities, 188
  corporate strategies, 212–215
Lindzen, Richard S., 90
Love Canal, Niagara, New York, 32, 59

Marini, Robert C., 238–239
Market-based incentives, 68–77, 96. *See also* Command-and-control regulations; Government policy
  categories of, 101
  corporate strategies, 220–222
  described, 68–70
  implementation difficulties, 75–77
  program advantages and disadvantages, 70–75
  technology and, 101–102
Markets. *See also* Environmental markets
  alliances and, 14
  asset-management perspective, sustainable development and, 51–52
  construction industry and, 3–4, 12
  economic models, 33–37
  energy generation, 50
  environmental movement and, 62–63
  globalization and, 6–7
  government policy and, 37–40, 58
Marks, David H., 226
Mass burning, solid waste technology and, 116
Megaprojects, decline in, 7
Microencapsulation, hazardous waste technology, 110
Military, hazardous waste disposal, 20, 22, 28. *See also* Department of Defense

INDEX

MIT Symposium on Global Environment and the Construction Industry, abstracts of presentations, 225–240
Moavenzadeh, Fred, 234–235, 238
Montreal Protocol, 91
Moomaw, William R., 227–228
Muir, John, 28
Municipalities, solid wastes and, 83

Nader, Ralph, 29
National Ambient Air Quality Standards (NAAQS), 80
National Cooperative Research Act (NCRA), 129
National Environmental Policy Act (NEPA) of 1969, 65, 77
National laboratories, technology, barriers to, 127–128
National Science Foundation, 128
Nelson, Gaylord, 29
New Deal, 64
Newsprint, 75
Niemczynowicz, Janusz, 52
NIMBY syndrome:
   environmental market opportunities, 148, 150, 151
   government policy and, 59
   solid waste management and, 113–114
Nongovernment organizations (NGOs), government policy and, 59
Non-point-source pollution, water pollution control, 82
Not in my back yard. *See* NIMBY syndrome
Nuclear energy sources:
   alternative fuels, 124
   France, 92

Occupational Safety and Health Act (OSHA) of 1970, 65, 214
Ocean pollution, 29
Office of Technology Assessment (OTA), 83
Oil embargo (1973), conservation and, xxii
Oxidation, hazardous waste technology, 111
Ozone depletion, 90–91. *See also* Global warming

Pacific Rim, environmental markets, 169–171
Partnerships. *See* Teamwork approach

Penalties. *See* Fines and penalties
Performance standards, air pollution, 100
Petrochemicals, environmental markets and, 163–164
Photovoltaic systems, alternative fuels, 124–125, 162
Physical separation, hazardous waste technology, 110–111
Politics. *See also* Government policy
   environmental legislation and, 31
   environmental markets, dynamics of, 136
   government policy and, 59–60
   technology, barriers to innovations required for, 125–131
Pollution control, environmental market structure, 137–138
Pollution fees/taxes, market-based incentives and, 71–72, 101. *See also* Taxation
Potentially responsible parties, hazardous waste cleanup, 84
Pozzolanic stabilization, hazardous waste technology, 109
Prepared-bed systems, hazardous waste technology, 106
Press, Frank, 19
Prevention, shift to, 2, 5
Project 88, recommendations of, 74–75, 101
Property rights, economic models, 35
Public opinion:
   citizen empowerment, described, 77–78
   environmental issues and, xx–xxii, 32
   environmental market demand creation, 134
   government policy and, 7–8, 58–59
Public relations, alliances and, 15
Public utilities. *See* Electric utilities; Energy utilities
Public utility commissions, market mechanisms and, 76
Public Utility Regulatory Policies Act (PURPA) of 1978, 50
Pyrolysis, hazardous waste technology, 107

Reagan, Ronald, 126
Recycling:
   bottle deposit laws, 74
   environmental market opportunities, 147, 148, 152–153
   solid waste technology and, 118–119
   trade-offs and, 60

Refractory-lined furnace, solid waste technology and, 116
Refuse-derived fuel systems, solid waste technology and, 116
Regulation. *See* Command-and-control regulations; Government policy
Reilly, William, 5, 12, 74
Remediation:
　environmental market growth, 140
　environmental market structure, 137
　of site, construction industry opportunities, 174
Research centers, technology, barriers to, 128
Research and development. *See also* Investment
　bioremediation and, 113
　construction industry opportunities
　　industry, 191
　　owners, 195–197
　energy generation, 49
　environmental markets and, 97–98
　sustainable development, barriers to, 19, 127
Resource Conservation and Recovery Act (RCRA) of 1976, 83, 84, 101, 143, 144, 145
Retrofitting, construction industry opportunities, 174
Right to Know laws, impact of, 77–78
Risk, corporate strategies, 212–215
Roosevelt, Theodore, 28
Rotary furnace, solid waste technology and, 117
Rotary kiln incinerator, hazardous waste technology, 108

Sanitary landfill, solid waste technology and, 114–115
Sarofim, Adel F., 235–236
Sewer/wastewater, environmental market opportunities, 154–155
Shurtz, James M., 168, 169
Site remediation, construction industry opportunities, 174
Social cost concept:
　alternative fuels versus fossil fuels, 162
　economic models, traditional, 35–37
　market-based incentives and, 68–69
Soil washing, hazardous waste technology, 110
Solidification/stabilization, hazardous waste technology, 109–110

Solid waste:
　command-and-control regulations and technology, 113–120
　environmental market opportunities, 147–153
　government policy and, 82–85
Solid Waste Disposal Act (SWDA) of 1965, 82–83
Solid Waste Disposal Act (SWDA) Amendments of 1976, 83
Source control:
　construction industry opportunities, 174
　environmental market structure, 138
　waste-management approach, 9–11
Stabilization. *See* Solidification/stabilization
Standards:
　air pollution control, 80, 99–100
　command-and-control regulations, 65–68
　water pollution control, 82
State regulation. *See also* Government policy
　air pollution control, 100
　technology, sustainable development barriers, 17
Stavins, Robert, 74
Stevenson-Sydler Technology Innovation Act of 1980, 127
Strategies. *See* Corporate strategies
Strelow, Roger, 143
Stussman, Howard B., 18–19, 232–233
Subsidies:
　alternative fuels, 162
　market forces and, 76–77
Sugimoto, Fumio, 235
Superfund program. *See also* Environmental Protection Agency (EPA)
　construction industry opportunities, 187–188
　described, 84–85
　environmental market opportunities, 143–144
　fines and penalties, 40
　hazardous wastes and, 84, 105
　historical perspective on, 32
　landfills and, 83
　risk and liabilities, 212–213
　sustainable development barriers, 17, 19–20
　teamwork approach in, 11, 12
Survey questionnaire, 176–178
　design of, 176–177
　respondent characteristics, 177–178
　text of, 242–279

INDEX

Sustainable development:
  asset-management perspective and, 51–52
  barriers to innovations required by, 16–20, 125–131
  construction industry and, xxvii–xxviii
  description of paradigm, 50–51
  economic growth and, 64
  goal of, and environmental issues, xxiv, 2
Symposium on Global Environment and the Construction Industry (MIT), abstracts of presentations, 225–240

Taxation, technology, barriers to, 127. *See also* Pollution fees/taxes
Teamwork approach, construction industry, 11–16
Technology:
  alliances and, 15
  alternative fuels and, 124–125
  command-and-control regulations, 67, 96, 99–100, 104
    hazardous waste, 105–113
    solid waste, 113–120
  communications limitations and, 21–23
  construction industry and, xxvi
  construction industry opportunities
    market perspectives (industry), 190–192
    market perspectives (owners), 195–197
  energy efficiency and, 120–124
  environmental market structure, 138, 140
  government policy and, 97–99, 103–105
  hazardous waste disposal and, 3–4
  market-based incentives and, 101–102
  sustainable development, barriers to innovations required for, 16–20, 125–131
  trade-offs and, 103
Technology-based standards, air pollution, 100–101
Telecommunications, construction industry and, 6–7
Tester, Jefferson W., 237–238
Thermal treatment, hazardous waste technology, 106–109
Thermoplastic binding, hazardous waste technology, 109–110
Third World. *See* Lesser developed countries
3M pollution prevention program, described, 42–45
Timber industry, subsidies and, 76
Times Beach, Missouri, 32

Timetables, compliance and, 104
Total Quality Management (TQM):
  AT&T pollution prevention program, 45–48
  environmental policy and, 40–41
Toxic wastes. *See* Hazardous waste disposal
Tradable permits, market-based incentives and, 71, 72–73, 101, 102
Trade offs, government policy and, 60
Transmission losses, electric utilities, 123
Transportation, of waste, interstate regulation of, 60
Tsongas, Paul, 126

United Nations Conference on Environment and Development (UNCED), 91–93, 159
United Nations Environmental Program (UNEP), 90
U.S. Forest Service, subsidies and, 76
Universities, technology, barriers to, 128

Valuation, of environment, corporate strategies, 218–219
Vienna Convention for the Protection of the Ozone Layer, 90–91
Vitrification, hazardous waste technology, 107, 108

Waste disposal, historical perspective on, 27
Waste-management approach:
  environmental market opportunities, 147–149
  environmental market structure, 138
  market-based incentives and, 71–72
  shift from, 2, 5
  source control for, 9–11
Waste minimization:
  environmental market opportunities, 147, 153–154
  solid waste technology and, 119–120
Waste-to-energy conversion, environmental market opportunities, 147, 148–149, 151
Waste transportation, interstate regulation of, 60
Wastewater. *See* Sewer/wastewater
Waterborne waste, government policy and, 81–82
Water Pollution Control Act of 1956, 81
Water Quality Act of 1965, 81–82
Water rights, selling of, 74
Water supply, environmental market opportunities, 155–156

Waterwall furnace, solid waste technology and, 117
Wet oxidation, hazardous waste technology, 107
Wet scrubber, solid waste technology and, 117
White, David C., 231–232
Whitelaw, Edward, 36
Wind power, alternative fuels, 124, 162
Wirth, Timothy E., 74